Manufacturing AI

Building the Data Foundation for the Next
Industrial Revolution

**Ryan
Andrew
Hill**

UMDA

Manufacturing AI: Building the Data Foundation for the Next Industrial Revolution

Non-Fiction Book

Disclaimer

Trademarks

ISBN-979-8-9993366-0-6

*To Kim, whose patience and support
made every page possible*

Contents

IIIIIIIIIIIIIIIIIIIIIIIIIIIIIIIIIIIIIIIIIIIIIIIIIIIIIIIIIIIIIIIIIIIIIIIIIIIIIIIIIIIIIIIIIIIIIIIIIIII

Manufacturing AI

Foreword

||||||||||||||||||||||||||||||||||||||||||||||||||||||||||||||||||||||||||||||||||||||||||||

Manufacturing is on the cusp of something extraordinary. After years, decades even, of wrestling with disconnected systems, patchwork integrations, and siloed data, our industry is finally asking deeper, more strategic questions. "How do we stop drowning in data and start using it? How do we design architectures not just for real-time visibility, but for intelligence? How do we build infrastructure that doesn't just support people, but enables AI?"

Much of this shift in thinking can be credited to the popularity and rally cry behind the Unified Namespace. The UNS has sparked curiosity. It has given engineers and architects a vocabulary to talk about data flow and context. And maybe most importantly, it's brought about discourse on a global level around manufacturing data outside of the walls of a product, a standard, or a platform. People are willing to share what works, admit what doesn't, and help each other move forward. And the timing couldn't be better.

Across the first three industrial revolutions, we saw steadily increasing productivity gains and shorter time spans between each transformation. Now, as we approach the conclusion of the fourth industrial revolution, the pattern has shifted. Over the past twenty-five years, we've had unprecedented potential driven by high-speed connectivity, dramatically lower computing and storage costs, and processing power beyond anything we could have dreamed two decades ago. Yet the productivity gains of earlier eras have not materialized. Why?

The answer lies in the gap between our potential and our readiness. Our industry has the tools, the concepts, and the community, but nearly all manufacturing operations still lack the foundational structure to turn vision into scalable and sustainable results. As beneficial as it is to publish data, build dashboards, and wire up event-driven systems, many teams are now discovering that they're still not ready for what comes next.

The age of agentic AI is here. Autonomous systems are learning to reason, coordinate, and act. Most operations aren't prepared, they are far from ready. They do not have the right architectures and have not set the correct foundation. As the fourth industrial revolution comes to a close, most of us have not achieved what it offered, we remain rooted in our Industry 3.0 frameworks and we are realizing far too late that we are behind.

For some of us, it has been easier to bury our heads and just run the plant. For others we have farmed out our strategies and handed over the reins to third parties and cloud solutions, for certainly they know what to do... right? The majority of us are simply not prepared to advance, we need a game plan, and we don't know where to turn.

I am glad you are reading these words. Ryan has written something rare. In this book he presents a pragmatic, scalable, and deeply considered framework for how manufacturing organizations can actually prepare for the future. It's called the Unified Manufacturing Data Architecture, or UMDA. And whether you're starting from scratch, extending an existing UNS, or migrating from legacy systems, this framework stands on its own.

Ryan has done something few others have. He's named the blind spots before they become disasters. He's documented patterns that most don't notice until it's too late. And he's offered a path forward that's both visionary and grounded. It is technical where it needs to be but never detached from real-world constraints.

This book is generous. It's insightful. And it's honest.

It's not just about architecting systems. It's about protecting investments and extending the shelf life of existing solutions. Accelerating outcomes.

Enabling intelligence. And most of all, avoiding the painful moment too many manufacturers know too well... when millions have been spent, and yet, the value still feels just out of reach.

So, if you're building. If you're fixing. If you're dreaming. Read this.

Then build better.

— *Jeff Knepper, President, Flow Software*

Preface

||||||||||||||||||||||||||||||||||||||||||||||||||||||||||||||||||||||||||||||||||||||||||||||||||

Manufacturing has always been a driving force behind progress. From the first mechanized production lines to the refined efficiency of lean methods, it has consistently evolved to meet new challenges and opportunities. Today, we stand at the edge of another major milestone, one powered by data and artificial intelligence.

Picture a factory where every machine and system works together seamlessly. AI agents predict equipment failures three weeks before they happen, automatically ordering replacement parts and scheduling maintenance during optimal production windows. Production lines detect quality variations in real-time and adjust parameters instantly, preventing defects rather than hoping to catch them downstream. Supply chains anticipate disruptions from weather patterns, geopolitical events, and market shifts, rerouting materials before delays impact production. This is no longer science fiction. These capabilities exist today in the most advanced manufacturing facilities around the world.

Yet for most manufacturers, this vision remains frustratingly out of reach. Many have invested in IoT devices, analytics platforms, and AI tools. They have teams of skilled engineers, dedicated data scientists, and leadership committed to digital transformation. Still, the results often fall short of the promise. The challenge lies not in their technology or talent. It comes down to something more fundamental, which is their approach to manufacturing data itself.

Consider what happens in a typical factory today. A vibration sensor on a critical piece of equipment starts recording unusual patterns. The maintenance system flags it as a potential issue. But that same sensor data, which could reveal correlations with product quality, sits isolated from the quality management system. The production planning tool, unaware of the developing equipment issue, continues to schedule high-priority orders for that same line. When the machine finally fails, the impact cascades through the factory because the data that could have prevented the crisis wasn't available.

This scenario plays out thousands of times daily across manufacturing facilities worldwide. Manufacturing generates data differently than any other industry. Factories incorporate equipment spanning decades of technological evolution. A single production line might combine mechanical systems from the 1980s, programmable controllers from the early 2000s, and IoT sensors installed last month. Each component speaks its own data language, follows its own timing, and operates within its own constraints.

Traditional data solutions were built for environments, such as finance or retail, with more predictable systems and uniform data flows. Manufacturing needs are different. It needs architectures that handle real-time data from thousands of controller tags, connect legacy equipment with modern analytics, and provide insights that support immediate action on the factory floor.

This book closes that gap by introducing the Unified Manufacturing Data Architecture (UMDA), a comprehensive framework specifically designed for the complexities of modern manufacturing. Unlike generic data platforms adapted for industrial use, the UMDA was built to handle the unique challenges of manufacturing data. This includes managing the mix of real-time and batch processing, integrating both legacy and modern systems, and maintaining the high levels of reliability and safety essential to industrial operations.

The UMDA differentiates itself because it recognizes that manufacturing data has physical, environmental, and regulatory constraints that are unique to each location. When a sensor measures temperature, that reading is tied to a specific piece of equipment, at a specific time, under specific operating conditions. The architecture must preserve and enhance these relationships rather than abstract them away. It must enable AI systems to understand what happened, why it happened, how it affects other processes, and what actions should follow.

Throughout this book, you'll discover how to build this foundation step by step. We'll start with creating a Common Data Model that brings order to disorganized data without forcing every system into rigid conformity. You'll learn how to implement data federation at the edge, processing information where it's generated while maintaining enterprise-wide coordination. We'll explore how to balance local control with global alignment, enabling facilities to adapt to their unique conditions while maintaining consistency across sites.

The later chapters will cover advanced capabilities like creating symphonies of interconnected systems that respond to changes with remarkable coordination, building AI-ready architectures that scale from pilot projects to enterprise-wide deployments, and preparing for future technologies like quantum computing that will impact manufacturing from the ground up. Each concept builds on the previous components, creating a comprehensive approach to manufacturing intelligence.

This shift represents a complete transformation in how manufacturers operate. Companies who embrace these new approaches are redefining what's possible, moving from reactive problem-solving to predictive optimization. By building integrated intelligence, they move beyond competing solely on efficiency and instead lead through innovation and adaptability.

The strategies in this book have shown success across diverse manufacturing environments from discrete component assembly, continuous chemical processing, batch pharmaceutical production, and complex aerospace manufacturing. They scale from single production lines to global manufacturing networks. Most importantly, they're designed to deliver value quickly, building momentum for broader change while establishing the foundation for long-term competitive advantage.

Whether you're a plant manager seeking to optimize daily operations, an engineer implementing advanced manufacturing technologies, an IT professional responsible for integration and security, or an executive planning digital transformation strategy, this book provides practical tools for immediate impact.

The concepts apply equally to incremental improvements and revolutionary overhauls, ensuring relevance regardless of where you are in your journey.

Manufacturing is evolving at unprecedented speed. New technologies emerge constantly. Customer demands grow more complex. Global competition intensifies. Supply chains face increasing volatility. The manufacturers who thrive in this environment will be those who turn data into their most powerful competitive weapon. They'll anticipate problems before they occur, adapt instantly to changing conditions, and discover opportunities for continuous innovation.

This book offers a blueprint to excel as an industry leader. By the final chapter, manufacturing data will be seen not as a technical hurdle but as a strategic asset. You will have practical frameworks to address today's urgent needs while building the skills and systems needed for tomorrow's opportunities. Most importantly, you will have the confidence to lead change that delivers measurable results while creating lasting impact.

The technology to transform manufacturing exists today. But even the most powerful AI solutions are only as useful as the data they consume. Success comes from adopting the right approach to put that power to work in a coordinated manner. This journey begins with recognizing that manufacturing has unique demands and requires custom architectures built to support them. Let's explore how to build that foundation and put its full potential into action.

Chapter 1

Starting the Data Journey to Smart Manufacturing

||||||||||||||||||||||||||||||||||||||||||||||||||||||||||||||||||||||||||||||||||

Introduction

The story of manufacturing is one of continuous evolution, where every breakthrough has paved the way for the next. Today's technologies are the product of centuries of innovation built on top of one another. To understand the future of manufacturing, we must first explore a bit of its past.

Every breakthrough in manufacturing history has been fundamentally about turning information into advantage. The craftsman's apprentice who learned to read the color of molten steel. The factory foreman who discovered that tracking defect patterns could prevent quality issues. The engineer who realized that sensor data could predict machine failures before they happened. Each generation has found new ways to capture, understand, and act on the signals hidden within their area of expertise.

Modern manufacturing is a living record of its own transformation. Walk the floor of any factory today and you'll see layers of progress side by side. Paper logbooks still track shifts while digital dashboards show real-time data. Analog gauges share space with IoT sensors. Skilled operators work hand in hand with AI tools. This mix represents how the industry evolves, one layer at a time, without losing the core pieces that continue to add value.

Understanding this journey matters because the challenges we face today echo those of the past. How do you balance efficiency with flexibility? How do you scale operations without losing control? How do you preserve critical knowledge while embracing change? Those who solved these problems before us left behind more than just machinery and processes. They left behind patterns and principles that still guide successful transformation today.

This chapter traces that evolution from the first attempts at systematic record-keeping through the digital revolution that brought computers into factories, to the current era where artificial intelligence turns data into real-time intelligence. You'll discover how each advancement built the foundation for what came next, creating the ecosystem that enables today's smart manufacturing technology.

More importantly, this history offers more than a fascinating backstory. It serves as a roadmap for navigating the next phase of manufacturing evolution. The same forces that drove past innovations are accelerating today's AI-led journey. Understanding how previous generations successfully adapted to technological change provides valuable insights for leading similar transformations in modern operations.

By the end of this chapter, you'll see your current challenges through the lens of manufacturing's greatest innovations.

From Manual to Digital Manufacturing

Early 20th-century factories were loud, chaotic places. Workers operated machines by hand while foremen walked the floor with clipboards, scribbling notes about production numbers and quality issues. Everything depended on human skill and experience passed down through apprenticeships.

This hands-on approach worked well for smaller operations. Skilled craftsmen took pride in their work and could adapt quickly to material changes or new specifications. Managers understood every aspect of production because they lived it daily.

But scale broke the system.

Building things like automobiles required thousands of precisely manufactured parts assembled in perfect sequence. Coordinating these tasks with pen and paper became impossible. When machines produced defective components, hours or days passed before anyone with authority learned about the problem. Meanwhile, bad parts piled up, creating waste and costly rework.

Filing cabinets overflowed with production reports, inventory logs, and shipping manifests. Clerks couldn't keep up with the paperwork volume. Finding historical records meant digging through stacks of documents and adding numbers by hand. Simple questions took hours to answer, leaving little time for strategic planning.

Technology offered hope. Programmable Logic Controllers (PLCs) changed manufacturing by introducing flexible control to industrial processes. Instead of rewiring systems for every production change, manufacturers could simply reprogram them. Factories shifted from rigid, single-purpose lines toward adaptable systems that responded quickly to new products and market demands.

When these controllers combined with Human-Machine Interfaces (HMIs) they enabled the revolutionary concept of real-time visibility. Operators could monitor machines while they worked, tracking production speeds, equipment status, and errors as they happened. Teams caught problems early, preventing small issues from becoming costly downtime.

Manufacturing Execution Systems (MES) took this further. These platforms centralized information from across the factory, creating comprehensive views that extended beyond individual machines. For the first time, decisions could be based on real-time, holistic understanding rather than isolated data points or delayed reports.

Data historians added another dimension. Manufacturers could store years of production information, unlocking insights into long-term trends, quality patterns, and equipment performance. Factories shifted from reactive problem solving to proactive strategies.

Cloud computing removed one of manufacturing's biggest roadblocks. With scalable storage and computing power, companies could finally make sense of massive datasets spread across multiple facilities. Local systems continued to manage real-time operations, while cloud platforms unlocked a broader view powering long-term planning, forecasting, and innovation.

Today's factories rely on this blend. Edge systems keep production fast and responsive. Cloud systems pull everything together, offering insights that guide smarter decisions across the enterprise.

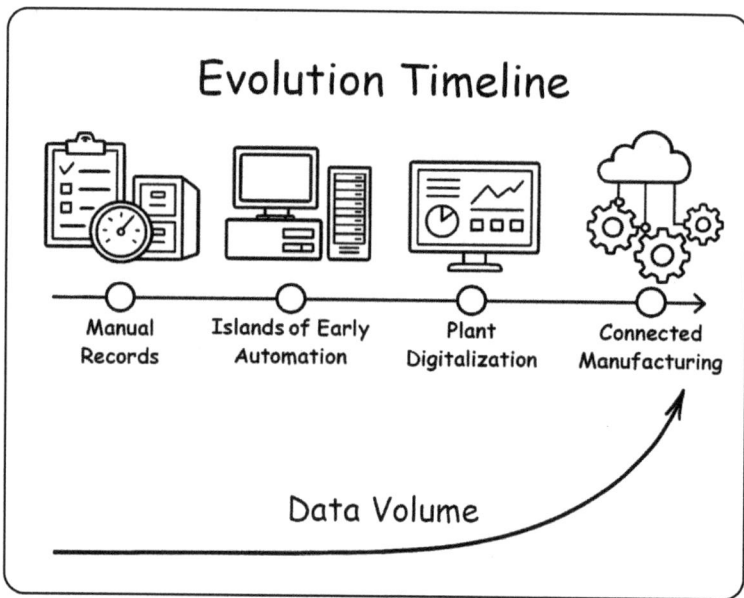

Evolution Timeline

Manual Records — Islands of Early Automation — Plant Digitalization — Connected Manufacturing

Data Volume

Figure 1-1 Technology Evolution Generates Increased Data

As manufacturing systems evolved to be more capable, they also became more complex. Success now depends on managing information across multiple layers, some urgent and others strategic. Real-time data keeps operations steady. Local systems monitor for signs of drift or wear and trigger action before small issues grow into bigger problems. This allows teams to respond in the moment rather than after the fact.

Centralized views from cloud systems bring everything together. Leadership can track inventory, energy use, and production status across multiple sites, making smarter decisions faster. They gain visibility into how changes in one area affect others and adjust before issues ripple through the system.

Longer-term trends add another layer. Patterns in cycle times, product quality, or equipment reliability often point to hidden opportunities. By combining historical insight with real-time data, teams can make precise improvements that lead to better outcomes.

Organizations must handle an extraordinary variety of information spanning from split-second sensor readings to multi-year performance trends. Managing this complexity requires structured approaches that balance immediate production needs with long-term strategic planning objectives. Early computerization efforts focused on specific tasks, yet these systems operated in isolation.

The Data Integration Challenge

Despite technological advances, a fundamental problem persisted. Factories collected more information than ever before, yet this data remained trapped in fragmented systems. The same technology that improved data collection created isolated pockets of information that couldn't easily be shared.

Data silos form naturally as organizations grow. Each department selects tools that excel at specific tasks. Production lines choose systems optimized for speed and throughput tracking. Quality departments invest in precision measurement and statistical analysis. Maintenance teams deploy specialized software for equipment monitoring and work order management. These focused solutions work well alone but often result in barriers between critical information streams.

Figure 1-2 Data Silos Built within Domains

The impact becomes clear during routine investigations. Quality engineers tracking down defects need production parameters from one system, equipment conditions from another, and material properties from a third. Each system requires different credentials, uses unique interfaces, and stores information in incompatible formats. Hours pass extracting and aligning data that should connect seamlessly. Investigations that should take minutes stretch into days.

Maintenance teams face similar challenges. Preventive maintenance often follows calendar schedules instead of actual equipment conditions because performance data remains separate from production results and quality records. Bearings might operate smoothly long after planned replacements or fail sooner than expected. Teams struggle to fine-tune maintenance timing without connected information linking equipment health to quality and efficiency.

Production planners work through even greater complexity. Sales forecasts live in spreadsheets. Inventory levels sit in warehouse systems. Equipment availability depends on maintenance schedules stored elsewhere. Material needs pull from separate procurement tools. Building reliable production plans means stitching together reports from scattered sources, often using information that's already outdated.

Technical hurdles add to these difficulties. Legacy machines generate information in formats modern tools can't read. Proprietary protocols trap data inside vendor specific systems. Even when connections are possible, inconsistent naming, measurement units, and update rates create confusion. Temperature readings labeled "TempSensor_01" in one system might appear as "Temperature_Line1_Zone3" elsewhere, with no built-in way to link them.

Organizational structures further reinforce these gaps. Departments shape their own metrics, processes, and priorities. Production focuses on throughput. Quality aims for zero defects. Maintenance seeks longer repair windows. Logistics needs predictable schedules. Teams optimize in isolation while overall performance suffers.

The ripple effects reach executive levels. Leaders making strategic decisions rely on reports that aggregate fragmented information into simplified views. They miss relationships and patterns that would emerge from integrated data. Successful efficiency improvements at one facility stay hidden from similar plants that could benefit. Investment decisions lack complete operational context needed for optimal allocation.

Predictive models miss early warning signs when they rely only on equipment data without understanding what's happening in production. Process improvements stall because no one sees how one shift affects the next. Supply chain decisions fall short when inventory numbers don't line up with what's actually happening on the floor.

It takes a toll on people too. Engineers waste hours pulling reports instead of driving improvements. Operators do their best with limited information, unaware of how small choices ripple downstream. Managers make calls based on a portion of the big picture, while always one step behind. Everyone knows the answers are in the data. The problem is getting to them.

Addressing siloed data is as much organizational as technical. They reflect how organizations evolved, adding systems and departments to meet growing needs without considering integration. Each silo made sense when created. The challenge comes when these logical divisions prevent organizations from operating as unified systems.

Modern manufacturing demands connected information. Quality outcomes depend on production parameters. Equipment reliability affects scheduling accuracy. Material properties influence process settings. These relationships exist whether systems recognize them or not. Success comes from making these natural connections visible and actionable.

Organizations that eliminate fragmentation realize new benefits. Problems that once took days to diagnose resolve in minutes when information flows freely. Insights invisible in isolated systems emerge clearly from integrated data. Best practices spread naturally when successes at one site become visible across the enterprise.

This design pattern sets the stage for advanced technologies. Machine learning thrives on rich, connected datasets. Predictive analytics need comprehensive views to generate accurate forecasts. Digital twins require synchronized information from multiple sources to create accurate representations.

The pressure builds as manufacturing becomes more complex. Every new system, sensor, or software tool adds another layer of disconnected data. Without a clear plan for integration, the flow of information slows down while complexity speeds up. Meanwhile, competitors that connect their systems move faster, respond smarter, and learn more with each cycle. The real decision is whether to keep working around fragmented data or to invest in the kind of integration that drives progress.

Difficulties with Data Complexity and Volume

Harmonizing data sources improves cross-system visibility, but it also introduces new challenges. As information flows across the enterprise, companies face a fresh problem. The flood of data that promised better visibility now threatens to overwhelm the very teams it was meant to help.

The scale of manufacturing information now defies traditional approaches. A single production line might generate millions of data points daily from sensors tracking temperature, pressure, and quality metrics. Business systems add layers of complexity with order details, inventory levels, and shipping schedules. Meanwhile, maintenance logs, operator notes, and compliance records create streams of unstructured information that resist easy categorization.

Figure 1-3 Disconnected Data Sources Lead to Confusion

Consider what happens when a quality engineer investigates a defect trend. The answer might lie in machine settings from three weeks ago, combined with humidity readings from yesterday, cross-referenced against material certificates from a supplier database. Finding these connections requires navigating multiple systems, each with its own format, update frequency, and access method. The result is hours of manual detective work to answer what should be simple questions.

The challenge goes far beyond sheer volume. Manufacturing information comes in many forms that rarely fit neatly together. Machine controllers work in milliseconds and binary signals. Enterprise systems focus on transactions and business timelines. Quality platforms collect precise measurements that only make sense in the right production context. Lab results might show up as PDFs, while sensors send continuous streams of values. Each system was built to solve a particular problem, leaving manufacturers with a jumble of formats.

Timing adds another layer of complexity. A spike in bearing vibration calls for action right away. Temperature trends might take months of information to reveal seasonal effects. Inventory updates could come by the hour, while financial systems work on monthly cycles. These different tempos create a puzzle where linking cause and effect becomes a struggle. When defects show up, teams can't easily tell if they were triggered by this morning's temperature shift or a material change from last week. Without synchronized timestamps and the right context, even straightforward questions become tough to answer.

The human element adds another dimension entirely. Experienced operators know their machines' quirks, but this knowledge lives in notebooks, shift logs, and memory. Maintenance teams track repairs in one system while engineering changes live in another. Sales forecasts that drive production planning exist separately from the capacity constraints they're meant to respect.

Organizations caught in this complexity face real losses. Production problems that should be simple to resolve drag into lengthy investigations. Managers make choices based on incomplete information because pulling together the full picture takes too long. All the while, important patterns that could drive improvements remain buried, waiting for someone with enough time and persistence to uncover them.

Storage costs are another concern. High-frequency sensor information can consume terabytes of space quickly, and cloud costs keep climbing. The real cost is what teams miss. Without a unified structure, it's hard to trace quality problems back to their root cause. Efficiency gains that could come from spotting patterns across equipment are missed. Failures go unpredicted until it's too late to avoid costly downtime. Best practices stay stuck at individual sites because information doesn't flow in a consistent way.

This uncertainty leaves organizations unsure how to move forward. Which information holds real value? What can safely be archived? How much history is enough? Without clear guidance, teams end up storing everything, creating massive digital piles where key insights get lost. Instead of enabling smarter decisions, this overload leads to hesitation and delays.

Within this challenge lies opportunity. Companies that master information complexity gain decisive advantages. They spot problems before competitors even know to look. They optimize processes others consider fixed. They make confident decisions while others debate incomplete information. The difference isn't in having more information, but in making it work together intelligently.

Leading manufacturers recognize information as a strategic asset requiring careful cultivation. They invest in architectures that bring order to chaos. They establish governance that ensures quality without stifling innovation. Most importantly, they create cultures where information literacy becomes as fundamental as safety training.

Manufacturing information will keep growing in both variety and volume. New technologies will introduce fresh streams of information. Regulations will require additional tracking. Customers will expect greater transparency. Organizations that build strong foundations now will be ready to take advantage of future innovations. Those that wait may find themselves overwhelmed by complexity while others move ahead.

The challenges are significant, but the rewards are just as great for those ready to meet them.

Insights in Action

At the end of each chapter, we'll step into the factory and visit Norman, a seasoned manufacturing veteran who's spent decades navigating the challenges of the shop floor. Alongside his colleague Rita, Norman faces the kinds of real-world problems manufacturers encounter every day. Together, they'll explore how the concepts and tools from each chapter can be applied to solve common challenges.

Norman leaned back in his creaky office chair, surrounded by stacks of yellowed maintenance logs and faded equipment manuals. He'd been working at this factory for over three decades, and sometimes it felt like nothing had changed.

But today was different.

Rita, the newly hired data scientist, burst into his office, tablet in hand and excitement in her eyes. "Norman, you won't believe what I've found!"

Norman raised an eyebrow. "Oh? Did you finally locate that missing shipment of ball bearings?"

Rita laughed. "Even better. I've been digging through some historical data, those old logs you've been keeping all these years. They're gold mines!"

She pulled up a chart on her tablet. "See this? By digitizing your maintenance records and combining them with our current sensor data, we can now predict equipment failures with up to 85% accuracy!"

Norman's eyes widened. He remembered the countless nights he'd spent scribbling down every squeak, groan, and hiccup of the machines and when the breakdowns occurred. "You mean all those years of record keeping weren't just busy work?"

"Far from it," Rita grinned. "Your meticulous data is the foundation of our new predictive maintenance model. It's like you've been training our AI system for decades without even knowing it!"

As Rita explained how they could use this historical data to optimize production schedules and reduce downtime, Norman felt a mix of pride and amazement. The future of manufacturing was more than just shiny new robots. It started with unleashing the potential hidden in decades of accumulated knowledge.

"Well, I'll be," Norman chuckled, glancing at his callused hands. "Looks like this old dog might have a few new tricks to learn after all."

Rita patted his shoulder. "Don't sell yourself short, Norman. You're learning new skills and helping shape the future of manufacturing. Now, how about we start digitizing those quality results next?"

As they began working, Norman realized that the journey from paper logs to AI insights required more than new technology. Success meant bridging past and future by converting workforce knowledge and legacy system data into machine learning.

Chapter 2

Foundations of a Common Data Model

||||||||||||||||||||||||||||||||||||||||||||||||||||||||||||||||||||||||||||||||||||||||||||||||||||||

Introduction

Managing fragmented data can feel more like solving puzzles than driving improvement. Maintenance records, production data, and quality reports all speak different languages, with their own formats and naming conventions. Rather than providing clear insights, this disjointed structure creates confusion. Signals that matter such as early warnings, shifts in performance, or hidden opportunities for savings, can get lost in the noise.

When data doesn't align, the systems built to use it fall short. Predictive models miss context. AI tools deliver half-formed answers. Valuable insights disappear not because they aren't there, but because the data doesn't come together in a clear manner. What should drive smart decisions turns into a source of friction.

This disconnection doesn't just slow things down, it exposes operations to risk. Downtime stretches out because no one sees the full picture. Costs rise when avoidable problems slip by unnoticed. As complexity grows, so does the gap between potential and reality. Without clean, connected data, even the best tools can't deliver what they promise.

Closing that gap starts with a shared understanding that spans every team, system, and function. A Common Data Model is used to reshape how companies handle information. Acting as a blueprint for organizing data, it creates a universal language across the business. From shop floor equipment to enterprise systems, every part of the operation uses the same terms and structure. A pressure spike during a production run, for example, can trigger different concerns, maintenance sees equipment wear, quality sees a risk to consistency, and energy teams see efficiency loss. A shared model ties those perspectives together so everyone is working from the same facts.

When data is consistent and accessible, collaboration becomes effortless. A production manager can instantly correlate schedules with real-time equipment performance. A quality engineer can analyze defect patterns across multiple facilities without wrestling with different data formats. A maintenance team can develop predictive strategies using comprehensive historical data.

This standardization becomes the foundation for everything that follows. Advanced AI systems need structured, reliable data to deliver reliable results. Without this shared framework, even the most sophisticated algorithms struggle to produce meaningful insights. With it in place, organizations can scale successful innovations across operations with confidence.

As factories embrace digital modernization, the lesson becomes clear that success doesn't depend on how much data you collect, but whether that data is truly usable. A well-designed Common Data Model turns information confusion into actionable intelligence, providing the foundation for innovation, efficiency, and adaptability in an AI-powered future.

The Need for a Common Data Model in Manufacturing

Manufacturing thrives on information. Equipment, sensors, operators, and systems generate a constant stream of data, offering immense potential for guiding improvement. Without a consistent structure to connect this information, valuable insights remain untapped and scattered.

Picture a food processing plant on a sweltering summer afternoon. The facility's massive ammonia chiller works overtime keeping production lines at safe temperatures. This single piece of equipment consumes more electricity than a small neighborhood, and energy costs spike during peak afternoon hours when the grid strains under demand.

The plant manager knows an opportunity exists. On days where temperatures will rise along with utility prices, the chiller could precool glycol storage tanks during cheap overnight hours, then coast through expensive afternoon rates. Simple in theory, but very difficult in practice.

The energy management system tracks kilowatt consumption in real-time but stores information using equipment codes in a vendor-specific format. Weather sensors predict heat waves that drive cooling demand, yet this streams through a completely different platform using metric units and UTC timestamps. The regional grid operator publishes hourly pricing in CSV files using their own standard format. Production schedules that determine cooling needs live in an enterprise system that identifies equipment differently than any other platform.

An engineer trying to optimize this system faces a puzzle with mismatched pieces. The chiller appears as "CHLR-01" in maintenance records, "Ammonia Refrigeration Unit A" in the energy system, and "Building 3 Cooling System" in production planning. Temperature readings come in Fahrenheit from older sensors but Celsius from newer devices. Timestamps scatter across time zones. Energy prices arrive in different formats depending on the market.

This disconnect leads to real, measurable costs. During a recent heat wave, the chiller ran at full power during peak hours when energy rates were ten times higher than overnight. The glycol tanks still held extra cooling capacity from the night before, but it went unused. The heat spike had been in the weather forecast for days. All the information needed to avoid that expense was there. The problem was, the systems couldn't work together to act on it.

Similar challenges multiply across facilities. Compressed air systems leak money through inefficient scheduling. Steam generation misses opportunities to leverage waste heat. Packaging lines run during peak demand periods. Each system optimizes its own priorities while missing the bigger picture.

The maintenance team provides another example. They schedule equipment service based on runtime hours and calendar intervals. But runtime alone doesn't tell the whole story. A compressor working against high ambient temperatures degrades faster than one running in cool conditions. Equipment pushing maximum capacity wears differently than systems operating at steady, moderate loads. Energy consumption patterns often signal developing problems before traditional metrics.

Without standardized information, maintenance misses these connections. They might service equipment that's running fine while missing units approaching failure. Unexpected breakdowns force reactive repairs during production runs, causing delays and quality issues. Energy waste from degraded equipment goes unnoticed because consumption information lives separately from maintenance records.

Quality teams discover their own version of this challenge. Product specifications depend on precise temperature control. Energy-saving initiatives that cycle equipment or adjust setpoints can affect product quality. Without integrated information showing both energy actions and quality outcomes, teams operate focused only on achieving narrow objectives. Cost savings in one area create expensive problems in another.

Small problems cascade into major disruptions. A minor scheduling conflict between production and maintenance becomes a breakdown during peak output. A missed opportunity to precool during cheap power rates forces expensive peak consumption. Quality variations traced back to energy-saving measures create customer complaints. Each incident erodes confidence in improvement initiatives.

With the current push to apply machine learning tools to manufacturing use cases, these issues become magnified. Solutions, even when successful on a single production line or facility, face significant barriers to scaling. The lack of consistent structures prevents these models from being applied across multiple sites or processes. What works in one location requires a complete rebuild to work elsewhere. This limits the potential to deliver broader organizational impact.

The challenges are even greater when attempting to use intelligent systems for higher-level corporate or supply chain decisions. Gathering information across facilities or partners becomes an enormous task when it's stored with varying definitions and formats. Advanced analytics thrive on large, well-organized datasets, while fragmented and inconsistent information undermines the ability to generate meaningful insights at scale.

Success in modern manufacturing requires breaking these obstacles. A Common Data Model provides the foundation by establishing shared definitions, consistent formats, and clear relationships between different types of information. Like a translator, it enables systems to share insights rather than hoarding them in isolation.

Benefits of Implementing a CDM

A properly structured Common Data Model creates the foundation for a smarter, more connected factory. It gives information a clear structure and shared language, turning raw data into a valuable resource. With this framework in place, teams collaborate more easily, decisions happen faster, and processes flow more smoothly.

One of the biggest advantages is how it simplifies communication between systems. Many manufacturing environments require constant manual intervention or custom fixes to align mismatched formats. A standardized framework eliminates these obstacles, allowing systems to share information effortlessly.

Collaboration improves dramatically. With a unified approach, teams across departments can easily access the information they need without translation or manual manipulation. Engineers refine production processes using insights from live information. Quality teams identify root causes of defects by connecting equipment performance with production outcomes. Managers adjust schedules with clear understanding of how changes impact resources, capacity, and delivery timelines.

Information quality is another area where the framework excels. By enforcing clear standards, it eliminates problems like duplicate entries, inconsistent measurements, or mismatched naming conventions. Reliable, high-quality information builds trust among teams, leading to better analyses and more confident decision making.

A structured approach positions manufacturers to embrace advanced technologies like real-time analytics, predictive maintenance, and intelligent optimization. It lays the groundwork for innovations that are only possible with consistent, reliable information. What starts as a solution to unify systems quickly becomes a platform for continuous growth.

With standardized structures in place, machine learning models can be trained more effectively and deployed with greater confidence. Once a solution works in one production line, it can be scaled quickly to others, reducing the time and effort needed to expand improvements across facilities.

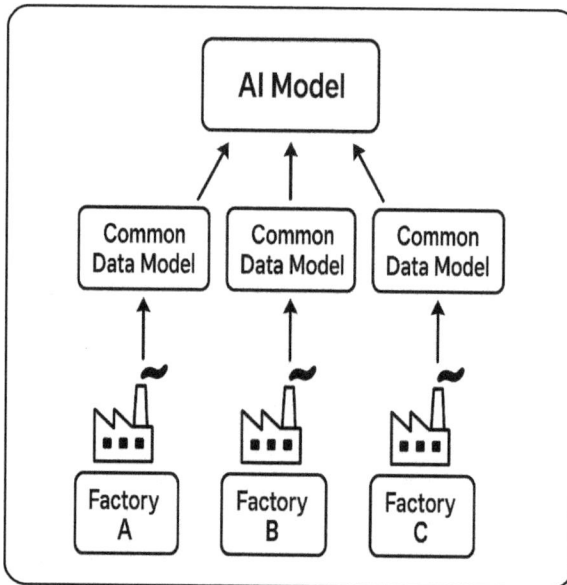

Figure 2–1 CDMs Enable Reusable AI Across Sites

Real-time analytics benefit from standardized structures. When information flows in a consistent format, it can be processed and analyzed instantly. Teams gain immediate visibility into production trends, quality issues, and equipment performance. Maintenance teams develop predictive strategies that use sensor readings and historical records to forecast failures and schedule repairs before downtime occurs.

Predictive modeling becomes far more powerful in this environment. By linking information from equipment, materials, and operations, organizations can build models that provide deep insights into complex processes. Quality predictions can take into account upstream factors, production variables, and inspection results to prevent defects before they happen.

The benefits extend beyond improving daily operations. Uniform patterns make sharing knowledge across sites seamless. Successes and innovations at one facility can be applied at others, multiplying their impact and accelerating progress. This scalability helps manufacturers stay agile and competitive in industries where the ability to adapt quickly becomes a differentiator for long-term success.

Supply chain operations thrive under a standardized approach. Consistent exchange between manufacturers, suppliers, and customers enables real-time tracking, automated processes, and more accurate forecasting. With everyone working from the same playbook, supply chains become more agile, efficient, and resilient to disruptions.

Business intelligence tools gain significant power when supported by uniform information. Dashboards and reporting systems can draw from across the organization without requiring manual configurations. Decision makers see the big picture clearly, uncover trends, and align strategies with long-term goals.

Even during major transitions like mergers or acquisitions, the benefits of standardization shine. Integrating new facilities or systems becomes faster and less complex when information already follows established patterns. This reduces delays and costs, helping businesses achieve the benefits of consolidation sooner.

This consistent approach reshapes how organizations innovate and grow. With a strong foundation, integrating new technologies, processes, or equipment becomes smoother and less resource intensive. Manufacturers can adopt new solutions confidently, maintaining workflow stability while staying ahead in a constantly evolving landscape.

Designing a CDM for Manufacturing

Creating an effective Common Data Model for manufacturing goes beyond meeting today's operational needs, it lays the groundwork for AI-driven innovation. A well structured model organizes data in a consistent way, while also accounting for differences that may exist across systems, processes, or sites. Furthermore, it brings clarity and structure, making data easier to use in both day-to-day operations and long-term planning.

The design process starts with understanding the unique challenges and priorities of the business. While technical tools are important, the foundation of a successful data model lies in capturing the insights of the people who know the operation best. Production managers, quality teams, maintenance crews, and IT specialists each bring essential perspectives. They know which data drives decisions, where existing systems fall short, and what improvements will create the most impact. Engaging these stakeholders early ensures the model reflects real-world needs while addressing gaps that limit performance.

A CDM designed with AI in mind opens the door to powerful analytics and predictive capabilities. Whether the goal is to optimize production schedules, predict maintenance needs, or improve product quality, the model must structure data in ways that help AI uncover insights and scale solutions across the operation. The data domains covered by each CDM will vary depending on the industry and business priorities, with each model shaped to capture the context and relationships most relevant to its area.

As an example, a Common Data Model for production operations captures the foundational elements of manufacturing such as equipment, materials, work orders, operators, and processes. These become dynamic entities, each playing a crucial role within the larger system. A machine, in this context, ties directly to production schedules, maintenance history, and performance metrics. These connections reveal the intricate relationships between individual components.

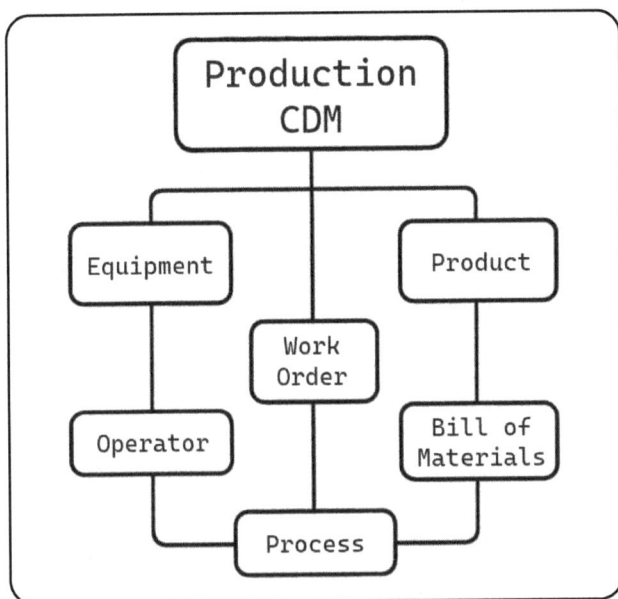

Figure 2-2 Production CDM Example Entities

Each entity in the model represents a vital aspect of manufacturing. Equipment entities go beyond basic identifiers like serial numbers. They also include operational parameters, maintenance schedules, and performance data, creating a rich profile of each machine's role. Material entities track attributes such as physical properties, lot numbers, and quality specifications, ensuring consistency across production and supply chain operations. Production orders serve as the bridge between customer demands and operational execution, linking schedules, resource allocation, and output measurements. Together, these entities build a cohesive, comprehensive view of manufacturing operations.

Equally important are the relationships between these entities. Manufacturing processes are deeply interconnected between upstream and downstream operations. Furthermore, a product is more than an end result. It reflects the raw materials used, the production methods applied, the quality checks performed, and the delivery dates required for customers.

The model must capture these relationships clearly to enable teams to trace issues, identify patterns, and improve processes. For instance, a single production order might involve multiple machines, materials, and operator roles, all linked to specific quality requirements. Capturing these interactions accurately, including scenarios like inline and offline quality checks with distinct timing or validation criteria, ensures the model reflects real-world complexity.

The depth of the model comes from its detailed attributes. Data points like temperature readings, material properties, and equipment tolerances aren't just numbers, but carry critical context. Units, precision requirements, and acceptable ranges ensure the data is universally understood across teams and systems. This consistency reduces errors by eliminating ambiguity and ensuring data remains clear, comparable, and actionable, regardless of its origin.

Standardized attributes enable seamless integration between systems, support advanced analytics, and enhance reporting capabilities. Anticipating future needs during the design process ensures the model remains flexible and valuable as technology and business needs evolve.

To make this wealth of information manageable, the model must be organized into logical layers. Factories can be structured hierarchically, starting with production lines, drilling down to individual machines, and ultimately reaching specific processes. This clear organization makes it easier to navigate complex operations while maintaining data accuracy and integrity. A well-defined hierarchy helps all stakeholders understand how different aspects of the operation connect and interact, providing a solid foundation for analysis.

Products and materials move through manufacturing operations in complex ways. A single product family might include many different models, each with its own variations. Materials flow through the factory from multiple suppliers, mixing and combining to create finished goods. Production plans change constantly, adapting to customer needs, equipment status, and material availability.

The challenge lies in tracking all these moving pieces. Teams need detailed information for specific tasks but also broad views of overall operations. They must know exactly what goes into each product while maintaining a clear picture of total inventory. A good data structure makes this possible by organizing information in logical layers, letting people quickly find the details they need without losing sight of the bigger picture.

Time is another essential consideration in the design process. Some systems generate data every second, while others update over hours or even days. Maintenance logs need to align with equipment performance records, and quality measurements must connect to specific moments in production. A strong data model accounts for these time-based relationships, ensuring insights remain clear and actionable.

Physical relationships within the factory also matter. The layout of equipment, material flow paths, and operator workstations all influence performance. A well-designed model incorporates these spatial relationships, helping teams optimize layouts and improve overall efficiency.

Structuring the Common Data Model using third normal form (3NF) is an effective way to ensure data consistency and reduce redundancy. By organizing data into logical groupings and eliminating unnecessary duplication, this normalized data makes the model easier to maintain and scale. It also improves data accuracy by ensuring that each piece of information has a single, clear place within the system. This structure simplifies updates and supports more reliable reporting, helping teams access accurate, up-to-date information when they need it.

Designing a Common Data Model goes beyond creating a technical structure. For it to succeed, people need to interact with it naturally and trust its outputs. Inaccurate data quickly destroys trust, and rebuilding that trust takes significant time and effort. Operators must find it intuitive to record data in standardized ways. Engineers should be able to locate information quickly for analysis. Managers should have straightforward access to metrics that inform decisions. A well-formed CDM makes these tasks easier, ensuring the model serves both practical needs and long-term goals.

When the structure works, it creates a stable foundation that supports continuous improvement. Teams can test new ideas knowing they have reliable data to track results. Changes to processes can be measured and understood across the entire operation. And when something works, it can scale more easily because the same standards apply everywhere.

Maintaining this consistency depends on data quality. Validation rules play a key role, ensuring that the information collected is accurate, reliable, and consistent. These rules set clear expectations such as requiring precise measurements, enforcing logical relationships between data points, or limiting values to specific ranges. A temperature reading might need to be recorded to a tenth of a degree, while material weights are checked to stay within machine capacity. These checks help ensure the data is trustworthy, turning raw numbers into insights that people and systems can rely on.

Industry standards like ISA-88 and ISA-95 provide a strong framework for shaping the model. These established frameworks provide a common language and approach for organizing data, based on decades of practical experience in manufacturing. Using these standards brings consistency across teams and systems, reduces confusion over terminology, and makes it easier to integrate with external partners or platforms. These standards aid in modeling critical concepts such as production scheduling, equipment information, process execution, and material tracking, enabling a unified approach across platforms. Mapping organizational data requirements to such frameworks establishes a scalable and robust model. We'll discuss more on how to apply these standards to data models later in the book.

Even with industry standards as a foundation, no single framework will fit every situation exactly. Each manufacturing operation has its nuances, requiring modifications or extensions to industry frameworks. Balancing these adjustments with established methodologies ensures the model addresses unique operational needs without losing its structured foundation.

A common challenge during design is reconciling terminology and data structures across different systems. Factories use a mix of platforms, each with its own definitions and conventions. For example, one system might call something a "work order," while another refers to it as a "job ticket." Industry standards serve as neutral references to harmonize these variations, ensuring consistent interpretation and communication across departments.

The design phase may highlight inconsistencies that have accumulated across departments over time. Teams might find they are tracking similar metrics with different terminology or levels of detail. Standardizing definitions, such as how cycle time or equipment performance is measured, eliminates confusion and ensures that everyone works from the same understanding. This clarity fosters collaboration and supports reliable, organization wide analysis.

A data model must grow with the organization. What works for one production line should scale easily to handle multiple facilities, different products, and new technologies. This means finding the right balance between standard-ization and flexibility. While core structures stay consistent, the model must adapt to local equipment, workflows, and processes.

Manufacturing operations evolve constantly. New products launch, equip-ment upgrades arrive, and processes change to meet market demands. The model should handle changes smoothly such as accommodating more real-time sensors to storing longer historical trends. Building in this flexibility from the start helps organizations adapt quickly to new opportunities and challenges.

Looking toward the future, the model must be ready to support emerging technologies. Artificial intelligence, augmented reality, and other innovations are reshaping manufacturing. A flexible data framework ensures that new data types, evolving relationships, and advanced analytical requirements can inte-grate smoothly. This adaptability prepares organizations to leverage new tools while maintaining a stable foundation.

A well-formed Common Data Model organizes complexity without sacri-ficing depth, making it easier for teams to analyze, share, and act on informa-tion.

Overcoming Challenges in CDM Implementation

Building and maintaining a Common Data Model is an ongoing process that requires careful attention. Manufacturing environments constantly change, with new equipment, information sources, and business priorities emerging continuously. For a framework to remain an asset rather than a hindrance, it must adapt without losing its core purpose. Success comes from balancing flex-ibility and stability, ensuring the model grows alongside organizational needs.

A significant challenge lies in implementing updates without disrupting operations. Adding new sources, refining structures, or integrating advanced features can unintentionally cause system failures if done haphazardly. Systems relying on the model for reporting, analytics, or process operations must con-tinue functioning smoothly, even as updates roll out. Redefining relationships between established elements can disrupt workflows if existing dependencies aren't considered. Ensuring older systems remain compatible with new updates is essential to balancing innovation with reliability.

To handle these updates effectively, ownership and accountability must be clearly defined. When changes are made in isolation by individual teams, the framework can become fragmented. A governance structure ensures updates are coordinated and aligned with broader organizational goals. Governance also fosters collaboration between teams, helping the model evolve in ways that benefit everyone.

As manufacturing operations expand, scalability becomes critical. A framework designed for a single production line can struggle to handle the demands of multiple facilities or diverse processes. Smart sensors and other advanced technologies generate vast amounts of information in formats the framework might not initially support. Successfully incorporating this new information without overwhelming the system requires careful planning and thoughtful design. A well-built framework should grow gracefully, integrating new types while remaining simple and efficient.

Clarity and accessibility are essential for making the model practical across the organization. Clear definitions and procedures ensure that information is recorded and interpreted consistently, whether it's a temperature reading, production metric, or maintenance record. Good documentation becomes the foundation for effective use, guiding teams in understanding and applying the framework. When everyone speaks the same language, collaboration improves and errors decrease.

Adaptability is another critical requirement. Manufacturing is never static. New equipment introduces different needs, quality standards evolve, and process improvements create new connections between existing information. An upgraded production line might require tracking entirely new parameters like equipment vibration or energy consumption. A flexible framework accommodates these additions without requiring major overhauls. This adaptability must also incorporate external changes. Shifts in customer demand, material suppliers, regulatory requirements, or market conditions can all create new information needs.

Organizations find success by taking a phased approach to implementation.

Starting small, with critical processes or focused improvement projects, allows teams to experiment with the framework in real-world conditions. Early wins provide confidence and momentum, demonstrating the model's value to stakeholders. These initial steps also offer opportunities to identify challenges and refine the framework, making it more robust before scaling to other parts of the organization.

A gradual rollout minimizes disruption and creates a testing ground for refining the model. Pilot projects offer opportunities to address potential issues, test functionality, and fine-tune processes. These smaller initiatives help resolve technical challenges while building support across the organization by demonstrating the model's benefits in action. As the framework demonstrates its value, organizations can expand its scope steadily, adding new departments, processes, or tools in a controlled way.

Shifting to a standardized framework changes familiar routines, which can bring uncertainty. People may wonder how these changes will affect their daily work and decision making. Early attention to these concerns through clear communication, training, and support helps build confidence and makes the transition smoother.

Engaging employees early fosters trust and collaboration. When stakeholders actively participate in defining the framework by providing input, sharing insights, and helping prioritize needs, they're more likely to feel invested in its success. Early involvement ensures the model reflects real-world requirements and reduces resistance to change.

Effective training eases the transition. Employees need to understand how the framework benefits their roles and the organization as a whole. Operators might learn how standardized, real-time information helps them address issues more efficiently, while managers might see how clearer insights improve production planning. Emphasizing these advantages helps shift perspectives from skepticism to enthusiasm.

Transparent communication throughout the implementation process builds confidence. Employees should know what to expect, including timelines, potential challenges, and the overarching goals of the initiative. Clear messaging about how changes will improve operations and benefit teams sets realistic expectations while reducing uncertainty.

The success of a Common Data Model depends on aligning its implementation with the organization's strategic goals. A balanced approach that meets current operational needs while remaining adaptable to future demands ensures the framework becomes a dynamic asset.

Integrating IoT and Time-Series Data into the CDM

As we've discussed, a well structured Common Data Model creates clarity from complexity by organizing information into an architecture that is reliable and accessible. But as connected devices flood operations with real-time data, the challenge grows. Sensors and machines generate continuous streams of time-stamped data, capturing the details of every movement, measurement, and cycle. Without a plan, this continuous stream of information becomes noise. To make sense of it, this data must fit into the broader data model.

Bringing IoT and time-series data into the CDM is essential. Without it, real-time operational data stays isolated. Teams lose the ability to connect what's happening on the production line to decisions made in planning, maintenance, or logistics. By integrating streaming data with the rest of the architecture, manufacturers gain the complete picture needed to act faster and more accurately.

Streaming data introduces unique challenges that require careful planning. The volume can be enormous, and the speed at which it arrives leaves little room for delay. As more devices connect, securing that data becomes more complicated. Furthermore, without clear context, even the most detailed data quickly loses its meaning. Addressing these issues is essential to ensure time-series data enhances the rest of the model, instead of adding complexity or risk.

The first challenge is volume. A single production line can produce millions of data points in a single day. Sensors constantly monitor temperatures, pressures, speeds, vibrations, and energy use, often recording measurements multiple times per second. When repeated across an entire operation, the scale quickly becomes overwhelming. Traditional databases, built for structured transactions, aren't designed to handle this kind of load.

The data architecture needs a strategy that fits the nature of time-series data. Storing every data point forever isn't practical or necessary. Tiered storage that adjusts data granularity over time is a better option. For example, collect high-frequency data for short-term analysis of a specific scenario. After that, aggregate the data into minute-by-minute averages, preserving important trends while reducing the footprint. For long-term insights, hourly or daily summaries offer a useful view without overwhelming the system. This approach keeps recent data detailed enough for immediate troubleshooting, while simplifying older data for long-term trends and AI model training.

Edge processing helps control volume at the source. Rather than sending every data point to a central system, devices process data locally, filtering out noise and sending only the most relevant events. If a machine runs within normal parameters, it reports summaries. When something goes wrong, it triggers alerts with detailed records. This reduces network traffic and storage needs, keeping the CDM focused on actionable data rather than raw streams.

Time-series databases, known as historians, also play a role. These systems are built for handling high-frequency, time-stamped data. They compress files efficiently and make it easy to retrieve data based on time ranges or conditions. When integrated into the broader CDM, they provide a specialized storage layer that holds raw and processed data without sacrificing performance.

Managing the volume of IoT data is only part of the challenge. The speed at which this data arrives creates another layer of complexity. Once data streams are flowing smoothly without overwhelming storage and systems, the focus shifts to how quickly that information can be processed and used.

In manufacturing, timing is everything. Traditional data concerning metrics might be collected at the end of a shift or the close of a production run. IoT data doesn't wait. It arrives in a constant stream, commonly in a matter of milliseconds, tracking conditions and performance in real-time. Systems that only manage transactional data can't keep up. To gain value from high-speed information, the data architecture must process and act on it the moment it arrives.

Stream processing is used to enable this. Instead of waiting for data to settle into a database before it can be analyzed, stream processing systems evaluate information as it moves through the pipeline. Imagine a production line where weight sensors measure each unit as it's packaged. If the fill level drifts outside specifications, adjustments are made automatically, preventing waste or under-filled products. This real-time correction reduces quality issues and ensures compliance without manual intervention.

By processing data in motion, operations shift from reactive to proactive. Small deviations are caught before they turn into bigger problems. Machine settings can be fine-tuned continuously based on actual conditions instead of historical reports. Downtime is reduced because maintenance decisions are informed by what's happening right now, not what happened yesterday.

Real-time capability opens up new opportunities, but it also raises security concerns. As more devices connect and stream data at high speed, every connection becomes a potential risk. Each sensor, controller, and gateway must be protected to keep the entire system secure. This is important for protecting production, quality, and even operator safety.

Security must be built into the data environment from the start. A zero-trust approach is a common strategy. Every device, system, and user is required to prove who they are and what they are allowed to do every time they connect. This reduces the risk of unauthorized access, whether it's accidental or intentional. It also prevents compromised devices from becoming a gateway to more critical systems.

Protecting data at every stage is essential. Sensor data should be encrypted from the moment it's collected, during transmission, and while stored. Access controls help ensure users see only what's relevant to their role. For example, a line operator may need to monitor temperature readings without access to system configurations. Role-based access strengthens security while making it easier for users to focus on the information that matters most.

Keeping devices up to date is another piece of the puzzle. Every connected device should be tracked with the associated firmware versions, patch levels, and network status. This inventory makes it easier to identify vulnerabilities before they become problems. Regular updates prevent known exploits from becoming threats. Treating IoT devices with the same discipline as any other digital asset makes security manageable and systematic, rather than reactive.

Strong security protects the integrity of the data itself. Without trusted data, it's impossible to rely on analytics or automated decisions. And without trust in those systems, teams are unlikely to adopt them fully. Securing IoT data streams ensures that insights, alerts, and actions are based on accurate information, strengthening confidence across operations.

Even with speed and security handled, there's still one more important challenge, context. Data only becomes useful when it's tied to the bigger picture. Time-series readings on their own are just numbers. They gain meaning when they are connected to the equipment generating them, the products being made, the orders they support, and the maintenance history of the machines involved.

A temperature reading means much more when it's clear which machine it came from, which product was running, what the expected limits were, and whether maintenance was recently performed. Linking this information together makes it possible to see what happened, understand why it occurred, and determine the next action. Without context, IoT data remains isolated. With context, it becomes actionable insight.

A well structured data model plays an important role here. It ties high-speed, high-volume IoT data and equipment readings to the broader operational landscape. Whether it's a vibration sensor on a motor or a flow meter in a production line, each data stream connects back to equipment records, work orders, quality checks, and customer shipments. This integration allows teams to track cause and effect, predict outcomes, and make better decisions before issues arise.

When time-series data is fully integrated into the larger data strategy, it moves beyond isolated readings and becomes part of a much bigger picture. This is the real value of IoT data. Not in the raw numbers, but in how they help manufacturers run smarter, more adaptive operations.

To get there, raw sensor measurements need to be aligned in the CDM so their value is tied to the equipment or process it is associated with along with the current batch or work order it may be linked to. This type of dynamic context needs to be built into the data model from the start. Every data point should carry details about the current process step, the product or batch in production, and any relevant quality standards. When this happens, even complex factory environments become easier to manage. Teams know what to focus on, and automated systems can respond more intelligently without constant oversight.

For AI and analytics tools, context can't be considered as optional. Machine learning models don't just need data, they need to understand the conditions under which that data was collected. Without context, AI struggles to find accurate patterns or deliver useful predictions. But when each sensor reading is tied to the full scope of operational information it creates a powerful foundation for more accurate insights.

Bringing IoT data into the Common Data Model makes AI solutions far more effective. Predictive maintenance models become sharper when they combine live sensor readings with past repairs, production volumes, and quality outcomes. Optimization algorithms can fine-tune processes in real-time, adjusting variables like energy use or raw material flow based on current conditions rather than preset rules.

This approach is creates a living digital representation of the entire operation. A system where real-time data, historical trends, and predictive insights all work together to support better outcomes. With this foundation in place, manufacturers are positioned to move beyond isolated improvements and toward a fully data-driven, AI-enabled operation.

Refining the CDM with Data Groupings

When a production problem needs fast answers, delays can be costly. A quality issue may require tracing back through machine settings, past maintenance records, and production logs to pinpoint the root cause. If the information is scattered across different systems or buried in spreadsheets, every minute spent searching means more defective products on the line and more time lost.

This scenario plays out in factories every day. A Common Data Model lays the foundation by organizing raw data. But even with a solid framework, making sense of the constant flow of information can be difficult.

Data groupings organize information into clear, logical categories. They bring order to complex datasets by clustering information by time (past, present, and projected future) or by purpose, such as operational use, analysis, or historical reference. They can also be grouped by functional domains, such as quality, maintenance, or production processes.

Poorly organized data can allow quality problems to continue unnoticed and make compliance reporting a last-minute scramble. As more manufacturers adopt AI-driven tools, these issues grow more consequential. AI models rely on consistent and clearly grouped data to deliver reliable insights. An algorithm designed to forecast equipment failures needs an easy way to connect current sensor readings with historical maintenance records and operating conditions.

Organized data groupings help both people and machines follow clear paths to the information they need. A maintenance technician troubleshooting an issue can quickly find current equipment data, past failure events, and standard operating settings. All of this is accessible because the relationships between these data points have been grouped logically.

The following diagram shows how these groupings work together to support efficient operations today and enable AI-driven insights for tomorrow.

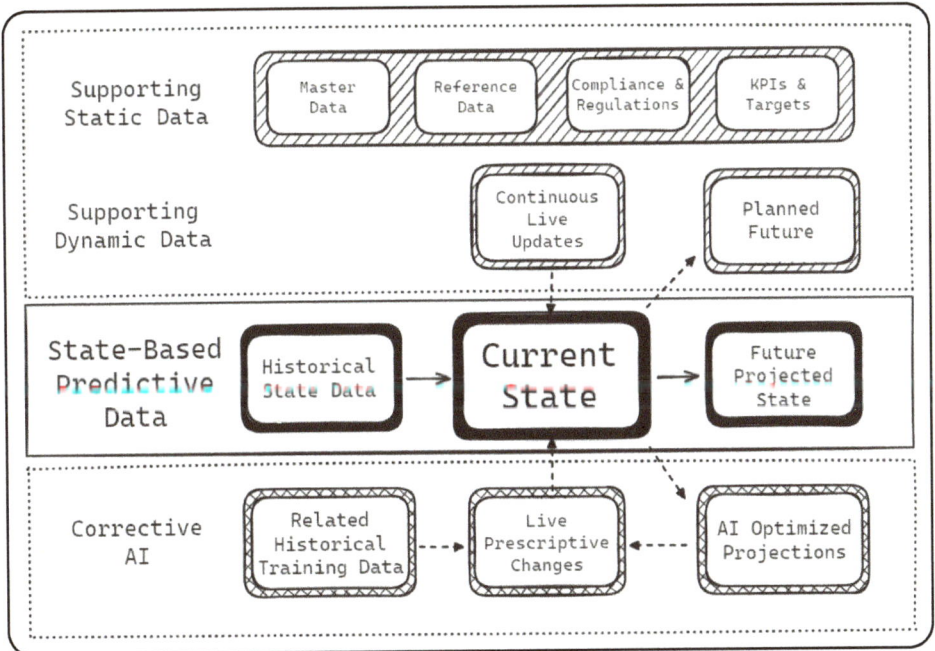

Figure 2-3 Leveraging AI for Production Optimization

This diagram shows how batches, work orders, equipment, production lines, or even entire sites can be analyzed by organizing data through logical groupings in the Common Data Model. It highlights how understanding the current state of any part of the operation creates a clear timeline by connecting past performance, present conditions, and future plans. This provides a complete view that can be applied to AI solutions to track progress and make informed corrections at every level to optimize production.

Historical State Data serves as the starting point of these groupings. This category collects all past records, offering a wealth of information to identify trends, understand long-term patterns, and benchmark performance. By analyzing this data, organizations can uncover recurring issues, measure the impact of previous decisions, and make informed choices to shape future strategies. It's the lens through which lessons from the past guide current actions.

Building on this foundation, the Current State grouping delivers real-time insights into the factory floor. This category provides a moment-by-moment picture of ongoing activities, capturing data on equipment performance, process metrics, and quality checks. It ensures that operators and managers always have a clear view of what's happening now, enabling rapid decision making to keep operations running efficiently.

The Future Projected State uses AI to predict what's likely to happen based on current trends and historical patterns. It provides a forward-looking view of operations, showing how things will unfold if no changes are made. This helps manufacturers anticipate potential issues, evaluate risks, and prepare for upcoming demands. By understanding where processes are headed, teams can make informed decisions about when and where to take action.

Supporting these state-based categories are additional data groupings designed to enrich their accuracy, provide real-time updates, and drive continuous improvement. Supporting Dynamic Data ensures the state-based groupings remain relevant and reliable. For example, Continuous Live Updates feed fresh information into the Current State category, seamlessly transitioning it into Historical Data while keeping the real-time snapshot accurate. This constant flow supports immediate responsiveness and ongoing adaptability.

The Planned Future grouping holds information about upcoming activities, such as production schedules, material deliveries, and maintenance plans.

Comparing this planned data against actual performance highlights deviations, helping teams identify areas for improvement and ensure alignment with strategic goals.

To get the most from these groupings, the Corrective AI category focuses on improvement and refinement. AI Optimized Projections represent the ideal future state that models work toward by analyzing current trends and historical data. This vision of better operations helps align production efforts with strategic targets. Live Prescriptive Changes provide specific recommendations for immediate improvements. These suggestions adjust processes and equipment settings in real-time, keeping operations on track with both planned and projected states.

Beneath these dynamic and temporal categories lies a stable foundation of Supporting Static Data, which provides essential reference points and contextual information. Master Data includes key business entities like customer profiles, supplier details, and product specifications, ensuring consistency and accuracy across all systems.

Reference Data, such as machine specifications, material properties, and product codes, establishes the standards for interpreting and using dynamic data effectively. Alongside this, Compliance & Regulations ensure operations meet legal and industry requirements, safeguarding the organization against compliance risks.

Finally, KPIs & Targets define the performance benchmarks for every facet of the operation. These metrics guide evaluations, highlight successes, and focus improvement efforts where they're most needed.

Organizing data into clear groupings turns the Common Data Model into more than a place to store information. It becomes a practical tool that brings clarity and drives action. Historical trends, real-time conditions, and predictive insights come together to give a complete picture of production. This structure lays the groundwork for advanced analytics and AI, making sure the system keeps up as needs change. By including placeholders for AI-generated forecasts and prescriptive adjustments, the model allows AI insights to flow directly into daily decisions. The result is a system where AI continuously learns and helps improve operations over time.

Data quality also sees substantial improvement through this structured approach. Clear categories and definitions make it easier to establish and maintain standards, reducing inconsistencies and enhancing the reliability of analytics.

Regulatory compliance benefits significantly from these groupings as well. A dedicated category for compliance and regulatory information ensures that critical data is always accessible and up to date. This supports audits, aligns operations with industry standards, and mitigates risks tied to non-compliance.

The integration of historical, real-time, and future oriented data offers a complete view of operations, bridging the tactical and strategic. Teams can address immediate challenges with confidence while also planning for long-term improvements. The ability to see how today's actions influence tomorrow's outcomes creates a proactive manufacturing environment ready to adapt to shifting demands and emerging technologies.

When implementing data groupings, it is most effective to begin with areas where improved organization will deliver immediate benefits. For many manufacturers, this involves prioritizing high impact processes or persistent problem areas. There is no need to restructure all data at once. A focused, incremental approach allows teams to demonstrate value early, gain practical experience, and build momentum for broader adoption over time.

Ensuring Data Quality and Governance in the CDM

A data model is only as valuable as the information it holds. Without accurate data even the best designed structure won't deliver the insights teams need. Reliable information is what turns a Common Data Model into a useful tool that supports daily decisions, long-term strategy, and AI-driven applications.

Strong data governance makes sure this doesn't happen. It sets the rules for how information is collected, validated, and maintained. It also defines who's responsible for managing it. In many ways, governance is the glue that holds the entire data framework together.

Data quality comes down to four measurable pillars that separate useful information from data that creates confusion. Accuracy means the data reflects reality. If a sensor reports a motor running at 2,000 RPM, that number must be correct. Even a small error can cause the wrong adjustment or hide a real issue that needs attention.

Completeness ensures nothing critical is left out. A product record missing a material spec leaves engineers guessing. A maintenance log with gaps can slow repairs or lead to mistakes. When data is incomplete people and AI are forced to fill in the blanks, which can lead to other problems.

Consistency keeps everyone on the same page. A part called a "bolt" in one system and a "fastener" in another creates confusion. Teams may waste time trying to confirm whether they're dealing with the same item. Consistency in naming, units of measure, and formatting avoids this problem.

Finally, timeliness ensures data is fresh. Yesterday's sensor readings won't help if a machine is overheating right now. Data that arrives too late leads to delays in action, and sometimes to bigger problems that could have been prevented.

Keeping data standards strong takes more than reminders or written guidelines. It calls for clear rules that everyone applies, backed by systems that make sure those rules stick. A plant might require temperature readings from equipment every 30 seconds, all using the same timestamp format. Procurement teams might follow set naming conventions for every new part, with required details like dimensions, supplier, and material grade. These practices build a common view of what good data means across the organization.

Implementing these standards requires systematic enforcement from the moment information is captured. Whether data comes from a sensor, a machine, or a human operator, it must meet the defined quality criteria before entering the Common Data Model. Validation rules act as the first line of defense, catching errors before they spread through the system. An application might flag a sensor reading that's wildly outside the normal range, signaling that equipment needs inspection. Or it could prevent a new part record from being saved if key specifications are missing. These automated checks stop small mistakes from creating larger problems downstream.

Data cleansing handles the more complex issues that arise from merging legacy systems or reconciling inconsistent processes. It identifies duplicate records, standardizes naming conventions, and resolves formatting differences. This step becomes critical when integrating data across production, quality, and maintenance systems. Clean, organized information moves smoothly through the data environment, enabling faster decisions and reducing operational errors.

Continuous monitoring ensures data quality remains high over time. Dashboards track error rates, highlight when certain data streams stop updating on schedule, and flag unusual patterns that might indicate system problems. If missing values spike suddenly, it signals that something in the data collection process needs attention. This ongoing vigilance helps teams spot and address issues before they can impact production. Monitoring also drives improvement by revealing recurring problems that might require process changes, additional operator training, or equipment updates.

Clear accountability makes sustained data quality possible. Each type of information needs a designated owner who understands the data and takes responsibility for its accuracy. Engineering departments might own part specifications, operations teams might oversee process data, and maintenance groups might manage equipment records. When ownership is clearly defined, expectations become concrete, and data quality becomes part of daily operations rather than an afterthought.

Regular audits provide additional protection and insight. Systematic reviews help identify recurring issues in production records, maintenance logs, or quality data. Beyond fixing immediate problems, audits can reveal root causes that point to needed system improvements. This analysis helps companies strengthen their data practices over time.

While quality standards make data trustworthy, governance frameworks define how that data is managed, accessed, and protected throughout its lifecycle. Governance policies establish who can access information, how it's used, and how it's secured. They provide structure for how data moves through the organization from creation to eventual disposal, ensuring that valuable information becomes a strategic asset rather than a risk.

Ownership and stewardship form the foundation of effective governance. Each area of the data model requires a clear owner who understands the information domain and accepts responsibility for its accuracy and appropriate use. When engineering departments manage part specifications, for example, any changes follow formal approval processes with clear documentation of what changed and why. This creates transparency, maintains accountability, and establishes a reliable audit trail.

Role-based access controls ensure teams work with exactly the data they need for their responsibilities. Engineers might control design specifications while production teams can view but not modify those same records. This approach protects sensitive information while keeping data well-structured.

Managing data lifecycle policies proves equally important. Every piece of information needs clear rules defining when it's collected, how long it remains in active use, and when it gets archived or removed. Detailed sensor data might stay in raw form for 30 days before being condensed into summaries for long-term storage. Legacy product information may need to remain accessible for warranty claims or regulatory requirements even after products leave active production. These policies keep systems efficient while preserving necessary historical records.

When governance and data quality management work together, they make the entire architecture stronger. They turn the Common Data Model into a solid base that supports advanced manufacturing intelligence. AI systems built on well-managed, high-quality data produce more accurate predictions, catch issues sooner, and suggest better improvements. This dependable base helps manufacturers move from basic data handling to smart, AI-driven factories. These practices ensure that as systems become more automated and connected, the information behind them stays reliable and consistent.

The Future of Common Data Models Enabling AI

The Common Data Model has proven its value by bringing consistency and structure to manufacturing data. It connects diverse sources, supports detailed analytics, and enables process improvements. Its future, however, holds even greater potential as it becomes a foundation for more intelligent and adaptive operations, driven by advancements in artificial intelligence.

The strength of this evolution lies in combining standardized data with emerging technology. While the CDM ensures data is organized and accessible, AI brings adaptability, identifying patterns and opportunities that are impossible to spot with traditional methods.

The CDM ensures the consistency and quality of the data used to train AI solutions. Reliable, organized information allows analytical tools to forecast trends, identify inefficiencies, and recommend adjustments. For example, a process engineer might use this to monitor how small changes in temperature or pressure affect product output. With AI trained on reliable data, the system can suggest optimal settings in real-time helping to maintain stability, improve yield, and reduce variability.

Looking ahead, these capabilities will become even more sophisticated. A future Production CDM might automatically detect when new sensor types appear on the factory floor and suggest exactly where they fit within the existing data structure. Instead of weeks of manual integration work, the system would propose mappings, validate data quality, and begin generating insights within hours. Quality engineers could deploy new inspection methods knowing the data would immediately integrate with existing trend analysis and predictive models.

The integration of data systems supports planning and quality control as well. Unified frameworks allow for comprehensive insights, whether optimizing production schedules to adapt to shifting demand or diagnosing the underlying causes of quality issues. By providing a complete view of the manufacturing process, these systems empower decision makers to act with confidence and precision.

The ability to analyze complex relationships between multiple variables is another step forward. Traditional methods miss subtle interdependencies between factors like machine settings, material quality, and environmental conditions. With the CDM enabling deeper analysis, manufacturers can use AI to uncover these interactions and make adjustments that improve performance. This enables AI to easily navigate the interconnected web of data, providing a comprehensive view of operations. This broad perspective allows manufacturers to optimize entire systems rather than focusing narrowly on individual components.

The integration of AI, IoT, and digital twins further amplifies this impact. Digital twins, virtual replicas of physical assets, are continuously updated with real-time data. This allows manufacturers to simulate "what-if" scenarios, predict outcomes, and refine strategies before implementing changes on the production floor. These simulations reduce risks, improve agility, and accelerate innovation. With AI analyzing and optimizing these virtual models, organizations can achieve a level of precision and foresight previously unattainable.

This collaboration aligns day-to-day decisions with broader goals, ensuring local optimizations roll up to measurable business outcomes. Decisions are informed by real-time insights, processes are optimized holistically, and workers are empowered with advanced tools that blend human expertise with machine precision.

As these technologies converge, they create new possibilities for manufacturing where data frameworks evolve alongside the operations they support. These frameworks move beyond static repositories to become adaptive systems shaped by AI, constantly learning to meet changing demands.

In this vision, AI doesn't merely use the data within the framework, it actively enhances and refines it. As we continue to incorporate technology in manufacturing environments, AI systems will soon autonomously integrate new data sources, interpreting their structure and significance while aligning them seamlessly with existing processes. By automating these tasks, organizations can dramatically reduce the manual effort of maintaining their data environments, freeing resources to focus on innovation and growth.

As systems grow more connected and data volumes increase, the ability to manage and trust that information becomes more important than ever. Data needs to be organized in a way that's clear, consistent, and usable across teams. It has to be accurate, current, and complete. Without this, even the most advanced technology will fall short. But with the right foundation, everything changes.

A unified approach brings together consistent data, time-series signals, and real-time insights into a single, straightforward model. Logical groupings add context. Governance builds confidence. Quality control ensures that the data entering the system is accurate enough to support dependable decision making.

Different roles interact with these data groupings in distinct ways that reflect their daily responsibilities. Operators work primarily with current state information, monitoring real-time equipment performance and quality metrics to keep production running smoothly. When AI suggests adjusting temperature based on humidity trends, operators see both current readings and projected outcomes in a single view.

Maintenance technicians take a different approach. They start with historical patterns to understand equipment degradation over months or years, then check current conditions for immediate issues, and rely on future projections for scheduling repairs. This workflow moves naturally from one data grouping to another, creating a complete picture of equipment health.

Plant managers need something broader. Their dashboards blend historical performance with current operations and future projections, enabling strategic decisions that balance immediate needs with long-term goals. They might see that Line 4 is currently performing well but historical trends suggest a maintenance window should be scheduled next month, while AI projections show optimal timing for a product changeover.

Of course, this shift introduces new challenges. Systems must be designed with security in mind. AI must be explainable and traceable. And as data becomes more central to operations, the culture around it needs to evolve too. People need to trust the tools they use and understand how to work alongside them. That trust comes from transparency and accountability.

The result is an environment where problems are addressed quickly, processes improve with clarity, and innovation can scale across the business. The Common Data Model gives users a strong foundation and the agility to respond to change. Both AI tools and operational teams can rely on the same clear, trusted data, making it easier to react with confidence when conditions shift.

From this point, the direction becomes clear. With structured models and organized groupings in place, the groundwork is set for processing data close to where it's generated while still keeping everything aligned across the enterprise. It brings consistency across sites and creates the conditions needed for intelligent agents to operate effectively, which we'll explore in more depth in the chapters ahead.

Insights in Action

Norman scratched his head as he stared out at the factory. "Rita, I've been here for all these years, and I still can't figure out why the heck our quality ratings always dip on Wednesdays."

Rita grinned, her tablet in hand. "Well, Norman, that's where our new Common Data Model comes in handy. It's like a universal translator for all our factory data."

Norman raised an eyebrow. "Universal translator? Like in those sci-fi shows?"

"Exactly!" Rita's eyes lit up. "See, our quality control system speaks one language, the production scheduler another, and don't get me started on the supplier database. Our CDM brings them all together in one happy data family."

Norman chuckled. "So, it's like the time we had that international company townhall and you set up a translation app so everyone could chat?"

"Bingo!" Rita nodded enthusiastically. "But instead of translating languages, it's translating data. And guess what? It just revealed why Wednesdays are problematic."

Norman's eyes widened. "You're pulling my leg."

Rita shook her head, still grinning. "Nope! The CDM let our AI connect dots we didn't even know existed. Every Tuesday afternoon, we do scheduled maintenance on the cooling system. The temperature takes about four hours to fully stabilize, which means our Wednesday morning shift starts with equipment that's still settling. The variation is tiny and within spec, but it's enough to affect the precision molding on our high-tolerance parts."

Norman slowly shook his head as this new understanding sunk in. "And we've been blaming the Wednesday crew all this time!"

Rita nodded. "Exactly. The AI spotted the correlation between maintenance logs, temperature sensors, and quality metrics across three different systems. Without the CDM making all that data speak the same language, we'd never have seen it."

Norman scratched his chin thoughtfully. "So, what do we do now? Change the maintenance schedule?"

"Not necessarily," Rita replied. "The AI suggests we could start the temperature recovery process earlier, add a preproduction calibration step on Wednesdays, or adjust our production sequence to run less sensitive parts first. We've got options now that we understand the real cause."

Norman nodded, impressed. "Alright, I'm sold on this CDM thing. But can it tell me why the coffee in the break room tastes like motor oil every other week?"

Rita laughed. "Sorry, Norman. It may offer standardized data visualization, but it's not a miracle worker... yet. But who knows? Maybe that's next on our innovation list!"

They both chuckled as they walked back to the control room, ready to tackle their Wednesday quality blues with their new data tool.

Chapter 3

Data Federation at the Edge

||||||||||||||||||||||||||||||||||||||||||||||||||||||||||||||||||||||||||||||||||||||||||||

Introduction

In high performing manufacturing environments, speed and clarity matter. The difference between leading and lagging may come down to milliseconds. A quality sensor detects a variation just as a part enters the next station. A machine begins to vibrate slightly outside normal parameters. Energy consumption spikes during peak production. These moments containing critical information can either drive immediate, actionable responses or get lost in the noise of disparate systems.

Responding quickly to this data is one of the hardest problems in manufacturing. How can facilities manage this flood of information without drowning in it? How can they ensure that insights reach the right place at the right time? And most importantly, how can they turn raw data into useful intelligence that directly increases KPIs?

The answer lies in applying strategies that change the way manufacturing data moves and works. Data federation brings information together from many sources without forcing it into a single location. Edge computing puts analysis and decision making close to where it can provide immediate impact. Contextualization adds the meaning needed for data to guide smart decisions.

When these capabilities coincide, they turn scattered signals into synchronized operations. Equipment responds in real-time to changing conditions. Systems adjust automatically to maintain efficiency. Supply chain data connects seamlessly with production metrics. The result is a continuous loop of insight that doesn't just monitor processes but actively improves them.

This represents a fundamental shift from isolated systems reacting to problems after they occur to a fully connected environment that can predict, adapt, and optimize continuously. Instead of waiting for centralized systems to process and redistribute information, intelligence happens where and when it matters most.

This chapter shows how these principles work within the Unified Manufacturing Data Architecture to create more intelligent, responsive, and resilient operations. From predictive maintenance and quality control to energy management and supply chain optimization, real-world examples demonstrate how the right data architecture improves manufacturing performance.

By the end of this chapter, you'll understand how data federation, edge computing, and contextualization work together to enable intelligent manufacturing.

Breaking Down Data Federation

We've established that manufacturing data sits in separate systems, each focused on its own job. Production details live in one place, inventory levels in another, and quality metrics somewhere else. These systems aren't designed to work together beyond what's needed for their specific tasks. That makes it hard to see the full picture of operations. Response times slow down, and valuable opportunities for improvement stay hidden. This fragmentation creates one of the biggest challenges in building a strong foundation for AI in manufacturing.

Data federation helps bridge this gap. Instead of managing individual buckets of data in scattered systems, federation brings everything together under a single interface, without necessarily moving or duplicating the data. Imagine a smart home hub that syncs the lights, thermostat, and security cameras into one app. Data federation achieves the same harmony for a factory, connecting disparate systems into one accessible and unified view.

Instead of transferring information into a central repository, federation establishes a virtual layer that allows data to be accessed and queried as if it were housed in one unified system. This virtual approach improves efficiency and reduces the delays, duplication, and potential errors that can come from physically moving or replicating data. Whether monitoring real-time production metrics, analyzing supply chain variability, or assessing equipment performance, data federation ensures that information remains organized, accessible, and ready for use across the enterprise.

By pulling data directly from its original source, this approach ensures that the information being used is always current. Outdated data can lead to costly mistakes, but with real-time access decisions can be made based on present conditions.

This approach works best when following a consistent framework. A Common Data Model should be used to ensure terminology and structure stay consistent. Federation makes that consistency usable by allowing teams and tools to access the data wherever it lives. Together, they form a foundation that keeps data connected and reliable across the organization. If a team wants to look at "cycle time," for example, the CDM ensures the definition is the same everywhere, while federation pulls that data from different sites into a single, clear view. It's a key part of Unified Manufacturing Data Architecture, removing the friction from cross-site analysis and cutting down on manual cleanup.

By linking systems and unifying data, federation eliminates the blind spots from distributed data, enabling AI and other technologies to operate with a full, accurate dataset. This succeeds because it adapts to what already exists. Companies don't need to replace current systems or force every dataset into a uniform database. Federation acts as a bridge, letting each system do what it does best while ensuring their data can work together. By connecting these different systems, federation helps companies make better decisions and achieve real operational improvements.

One of the most significant advantages of this approach lies in how it contrasts with traditional methods like data warehousing. Traditional data systems relied on physically moving or copying data into a centralized location, which could be effective but brought challenges. Warehousing required ongoing maintenance, constant syncing, and significant resources to ensure data stayed up-to-date. Any delays in synchronization could result in incomplete or outdated insights, a risk no manufacturing operation can afford.

By accessing data directly at its source, this modern approach eliminates the need for duplication and reduces the overhead associated with maintaining centralized repositories. Instead, a virtual layer enables real-time analysis, keeping up with the ever-changing conditions on the shop floor.

Speed alone doesn't deliver lasting results. Trust in the data is just as critical. When information comes directly from its original source, discrepancies are reduced and everyone relies on the same set of facts. Outdated reports and conflicting interpretations no longer get in the way. Whether fine-tuning production on the floor or shaping strategy in the boardroom, all stakeholders work from a shared, reliable foundation that strengthens confidence in every decision.

Linking previously isolated systems is among the most impactful steps in unifying data sources. In factory environments, vital information remains locked within specialized platforms or isolated departments. Production metrics might be confined to manufacturing systems, while financial details are restricted to ERP platforms. Creating bridges between these systems allows for a unified data ecosystem, enabling new insights to be revealed.

This approach allows teams to spend less time hunting for information across systems and gain immediate access to what they need. Advanced analytics and automated systems can access unified information directly, removing delays that come from navigating multiple disconnected platforms.

Achieving this level of integration requires both the right tools and a supportive organizational culture. On the technical side, robust solutions such as data virtualization platforms, API gateways, and middleware are essential to manage high data volumes efficiently. These tools ensure seamless connectivity across diverse environments, from IoT sensors on the shop floor to cloud-based supply chain applications. Selecting tools that align with a company's unique data landscape is critical to creating a system that works without disrupting plant operations.

Equally important is fostering a culture of collaboration and shared ownership of information. Many teams are accustomed to working within siloed systems and protecting their data. Shifting this mindset toward shared access and integration requires clear communication about the benefits, alongside training to help employees effectively use new solutions. When teams understand how easy access to integrated data can enhance their roles, the tools naturally become part of daily workflows.

For manufacturers that embrace this approach, the rewards are profound. Greater data accessibility leads to quicker responses to change, more informed decision making, and the ability to adapt with agility. This integration strengthens daily operations while creating the foundation for long-term innovation. Digital twins, advanced analytics, and AI systems perform best when backed by a unified, reliable data framework.

By turning fragmented systems into a connected, cohesive resource, manufacturers gain far more than operational efficiencies. They turn data into a valuable asset that drives improvement throughout the company.

The Role of Edge Computing in Manufacturing

As AI use cases trend closer and closer to the shop floor, the demand for real-time data is greater than ever. While data federation helps unify information from across operations, edge computing focuses on processing that data exactly where it's needed, at the source. Together, these technologies empower factories to move faster and respond smarter to production events.

Edge computing shifts data processing closer to the factory floor, where it's generated. In traditional setups, data from machines and sensors travels to a centralized server or cloud platform for analysis. While this method works for many use cases, it struggles with the speed required for split-second decisions.

The sheer volume of data in today's factories can also strain networks, slowing down critical responses.

By distributing processing power, edge computing keeps data local. Instead of waiting for instructions from a distant server, machines and devices can analyze and act on information immediately. This reduces delays and makes it possible to solve problems as they happen.

Consider a smart assembly line that monitors product quality in real-time. Sensors might detect a defect forming on a part as it's being assembled. With edge computing, the system can identify the issue, adjust machine settings, or even pause production right away.

This concept mirrors the way modern smartphones handle tasks directly on the device. Whether capturing a photo, using GPS, or sending a message, the device processes much of the activity locally rather than relying on remote servers. In the same way, edge computing brings localized intelligence to manufacturing environments, enabling systems to process data closer to its source for greater speed and efficiency.

With edge computing, manufacturing equipment becomes smarter and more autonomous. Machines can predict maintenance needs by analyzing performance data in real-time. Robots on the line can adjust their operations in response to environmental changes, such as temperature or humidity shifts. Even complex processes, like balancing production schedules with inventory levels, can be managed on the fly, minimizing the need for centralized oversight.

This localized processing also eases the burden on broader data systems. Instead of flooding a central server or cloud with raw data, edge devices filter and process the most critical insights.

Edge computing goes beyond improving individual machines. By enabling production lines, sensors, and robotics to work in harmony, it creates a real-time environment where decisions happen faster and smarter. This shift is changing how manufacturers use data and approach the future of leveraging AI across the enterprise.

At the heart of this infrastructure pattern is the need to overcome latency. In high-speed environments where decisions must be made in milliseconds, even slight delays can lead to disruptions, inefficiencies, or costly downtime. Traditionally, data has traveled between the site and data centers for processing, but this back-and-forth creates a bottleneck that no factory can afford.

Bandwidth is another concern this addresses. Factories are continually rolling out more IoT devices and sensors, each generating massive streams of data. Sending all this raw information to a central location can quickly overwhelm network capacity. This leads to bottlenecks, slow responses, and diminished efficiency. By shifting data processing closer to the source, edge computing eliminates these limitations. It allows decisions to be made locally, significantly reducing latency and optimizing the use of network resources.

To achieve this, edge devices such as gateways, industrial PCs, or smart sensors, are equipped with their own computing power. These devices are strategically placed throughout the factory to gather data directly from machines and sensors. Instead of simply forwarding raw data to a central server, they process it on-site, performing analysis and making decisions in real-time.

Reliability also gets a boost from edge computing. When systems rely entirely on centralized servers, a disruption in the network can bring operations to a halt. Computing at the site eliminates this dependency by enabling each edge device to operate autonomously. Even if external connections are lost, the devices can continue to function, maintaining production and ensuring critical decisions are still made.

The potential doesn't stop at operational efficiency. Edge computing opens the door to real-time analytics and advanced applications. With computing power distributed across the factory, manufacturers can now leverage AI and machine learning for immediate insights and actions. Quality control systems can spot defects the moment they appear, triggering on-the-fly adjustments to processes. Predictive maintenance algorithms can analyze performance data continuously, flagging early signs of wear and scheduling repairs before downtime occurs.

Consider a scenario where a sensor detects a subtle vibration in a piece of equipment. Without edge computing, this data might be logged, sent to a central system, and analyzed long after the issue has escalated. With edge devices in place, the vibration data is processed immediately, prompting the machine to slow down or alerting a technician before the problem worsens.

This approach creates an interconnected, intelligent factory floor. Machines, robots, and sensors work together seamlessly, analyzing their surroundings and coordinating their actions.

Implementing edge computing comes with its challenges, requiring more than just new hardware. It demands a shift from centralized IT systems to a distributed environment, where data is processed closer to its source. This shift introduces complexities in how devices are managed, secured, and governed. Each edge device becomes a critical node in the network, requiring monitoring, software updates, and alignment with operational needs.

Security also becomes a top concern. With systems distributed across a wide network, the risk of breaches increases. Ensuring every device is secure while maintaining compliance with industry standards is essential. These challenges require careful planning and robust strategies to ensure the system operates reliably and safely.

For manufacturers willing to take on these challenges, the benefits of edge computing are undeniable. Faster responses to operational issues, more reliable systems, and smarter, more adaptive processes redefine what's possible. One example of this was recently demonstrated by an industrial products manufacturer that produces high-pressure braided hoses. In this industry, even minor defects can have serious implications for safety and performance, making quality control a top priority. To address this, the company adopted edge computing combined with machine learning to modernize their processes.

High-resolution cameras were installed along the production line to continuously monitor the products at the equipment as they were manufactured. Each camera was connected to an industrial PC equipped with real-time data processing capabilities. These edge devices ran machine learning algorithms trained to identify defects with remarkable speed and accuracy.

When a defect was detected, the system immediately halted the production line, allowing operators to address the issue on the spot. This proactive approach prevented further waste and ensured defective products never reached customers. By catching problems as they occurred, the company saved both time and material while maintaining their commitment to quality and safety.

What made this solution particularly effective was its ability to adapt and improve. The machine learning algorithm was initially trained using historical defect data but continued to evolve with inputs from operators and data collected across multiple facilities. As it encountered new types of defects or environmental variations, like changes in lighting or seasonal conditions, the system adapted, ensuring consistent performance.

To manage this system across various locations, the company implemented a hybrid edge-cloud architecture. Edge devices handled real-time analysis and decision making, while the cloud served as a central hub for managing and updating the machine learning models. Updates were rigorously tested in the cloud before being deployed to the edge devices, ensuring reliability and consistency across all sites.

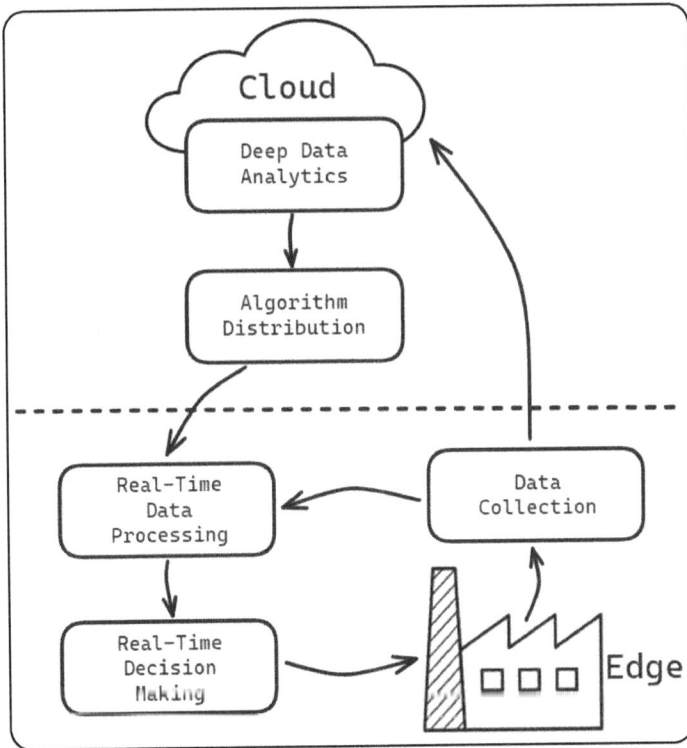

Figure 3-1 Edge / Cloud Distributed AI Cycle

The impact of this solution was significant. Real-time defect detection reduced waste by stopping issues before they escalated. The company enhanced product quality, safeguarded its reputation, and avoided liability risks. By transmitting only relevant information, such as confirmed defect images, to the cloud, they optimized bandwidth usage and avoided overloading their network.

This intelligent system solved immediate quality control problems while creating a solution that improved over time. Each facility benefited from shared knowledge across the network, building a system that continuously adapted to new demands. By bringing computing power directly to the production line, the company achieved speed, precision, and flexibility that traditional systems couldn't match.

This example shows how edge computing, combined with machine learning and cloud integration, can reshape manufacturing. The aim is to build AI models that solve problems and grow smarter and more efficient as they learn.

By enabling localized decision making and blending it with centralized coordination, edge computing helps manufacturers balance speed, precision, and scalability. This approach applies to far more than a single application or process. The principles extend throughout facilities, creating new opportunities for growth.

The Importance of Contextualization

While data federation and edge computing provide some foundation for the data architecture, they don't have a lot of value without providing meaning to the data. To fully realize this potential, raw information must be enriched with context by adding layers of detail to transform isolated data points into actionable insights. This process, known as contextualization, connects data points to the bigger picture enabling AI and other advanced tools to turn raw values into a full picture that can drive action.

Context gives data its significance. A single data point such as a vibration reading, a pressure value, or a machine speed has limited value on its own. This works by weaving together information from many different systems, giving each data point the meaning it needs to be useful. Production schedules from enterprise platforms provide a view into what should be happening at any given moment, while Manufacturing Execution Systems (MES) offer a live view of what's actually taking place on the shop floor. Historical performance data adds another layer, helping teams compare current trends to past benchmarks. Information about equipment gives added clarity when analyzing production or diagnosing issues. Even environmental factors like temperature and humidity play a role, especially when working with sensitive materials.

Understanding data in isolation is rarely enough. A sensor reading of 250°F doesn't mean much on its own, it's just a number. But connect that number to the machine it came from, the product being run, the operator on shift, and the environmental conditions at the time, and it starts to tell a story. That story becomes even more useful when compared to past patterns. Has this reading happened before? What was the outcome? Was it linked to a known issue, or is it typical for this kind of run?

Context shows whether a temperature sensor's reading is within expected limits or signals an issue. It clarifies whether a machine is under normal stress, something has changed in operations, or it's just being cleaned. Without this clarity, teams waste time reacting to symptoms instead of understanding the root cause.

AI systems face the same challenge. They can't learn or recommend improvements if they don't understand what the data represents. A temperature spike might look like a problem, but without knowing the production conditions, the materials in use, or whether a tool change just happened, the system can't make useful recommendations.

Contextualization solves this. It links sensor data with everything around it such as production orders, equipment state, maintenance history, environmental factors, operator inputs, and more. It pulls in historical records to show patterns, builds relationships between data points, and frames each number in the setting that impacted it. With this full picture, AI tools can move from detection to explanation. They can flag subtle trends, suggest targeted actions, and even anticipate future disruptions before they happen.

For example, a temperature sensor attached to a critical piece of equipment may report the same value across two different runs. But what that value means can change drastically depending on whether the machine is idling, running at full load, or operating under tight tolerances. If the material batch has changed, the ambient temperature shifted, or a scheduled calibration was skipped, the meaning of that same 250°F changes.

The following diagram shows how this comes together, illustrating how local and remote contextual data sources combine with IoT sensor data to build a more meaningful, connected data foundation.

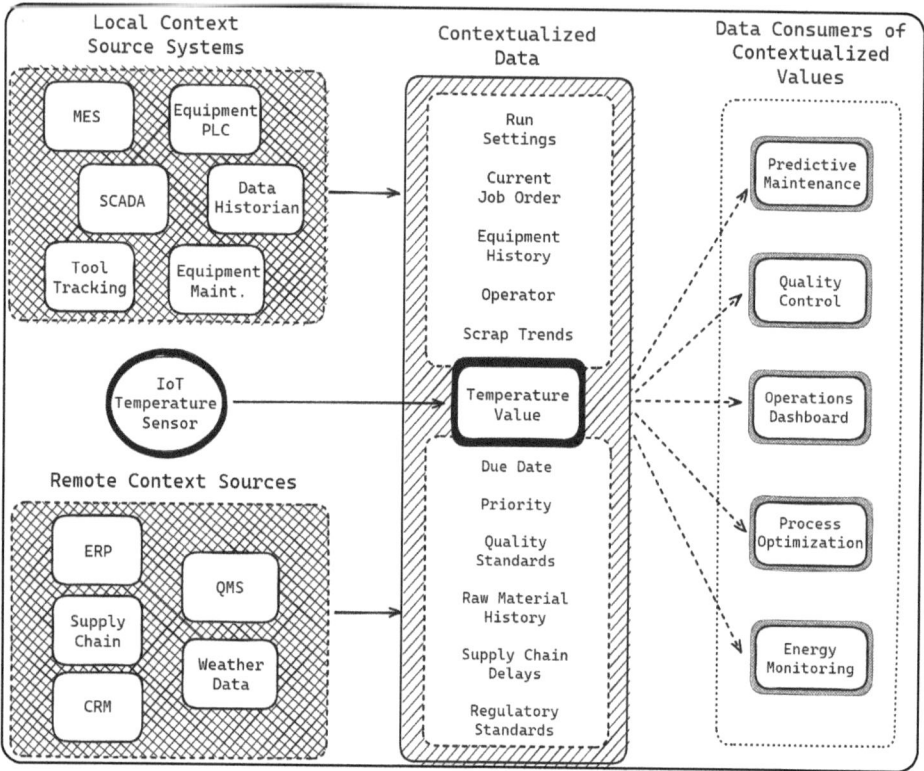

Figure 3-2 Data Flow of a Contextualized Sensor Value

The diagram highlights how a single data point gains meaning when connected to its surrounding context. A temperature reading from a sensor means nothing on its own. It becomes more valuable when combined with data from local systems like MES and SCADA, which provide current job details, equipment status, and recent maintenance records. Remote platforms such as ERP and QMS add supplier information, material specifications, and compliance requirements. Together, these layers create a clear picture of what's happening and why it matters.

This structured view supports a wide range of advanced applications.

Predictive tools can identify subtle shifts like a vibration increase paired with rising temperature, which signals an approaching equipment failure. Quality systems can trace a defect back to its root cause by analyzing raw material attributes, environmental conditions, and machine parameters in combination. These insights don't come from any one system alone. They emerge from the relationships between systems, brought together through contextualization.

By enriching the data environment, contextualization also strengthens collaboration across teams. Engineers gain visibility into how process settings affect quality. Maintenance teams see the connection between equipment wear and production outcomes. Operational leaders can make decisions with confidence, knowing the data shows what happened and why. Everyone works from the same foundation, using shared information to solve problems faster and more effectively.

With the addition of contextualization, the UMDA allows AI to understand the significance of data within the full operational environment. This ability to weave data into a coherent story is what enables AI systems to deliver more accurate analyses and informed insights tailored to the unique complexities of the factory.

Take the earlier example of using edge computing for quality control in industrial hose manufacturing. The high-resolution cameras installed on the production line don't just snap images, they enhance the data with contextualized information. Each photo can now be enriched with details like the specific production line where it was taken, the material being processed, the ambient factory conditions at the time, and the maintenance history of the equipment involved. This added information allows the AI system to move beyond defect detection. It can identify potential root causes, such as a correlation between temperature fluctuations and product inconsistencies. Then recommend preventive measures to avoid future issues.

Contextualized data extends far beyond the shop floor. Once integrated into the federated system, its value grows significantly. Quality issues detected in one system can be analyzed alongside production parameters from another, revealing hidden patterns. Supplier performance data might show connections between material quality and defect rates across multiple plants. By viewing these connections together, contextualized data provides insights that isolated information never could.

This ability to connect the dots is particularly valuable for predictive and prescriptive analytics. When AI systems have access to the full holistic view of past events, they can make more accurate predictions about future outcomes.

Contextualization positions AI as a strategic partner in manufacturing. With a harmonized understanding of the entire operation, AI systems can uncover opportunities for lean initiatives, recommend process changes, and guide product development.

Achieving effective contextualization within the UMDA requires thoughtful design and execution, and a unified data framework from the Common Data Model plays a key role in supporting this effort. By standardizing how data is defined, linked, and structured, it ensures that contextual information is captured in a clear and consistent way. This practice helps manage the complexities of integrating data from diverse systems, which is critical for advanced AI applications.

The first challenge is identifying which details matter most. Not every data point needs enrichment with layers of metadata or historical information. To begin, pinpoint the connections that add the most value. This could include tying machine settings to quality metrics, linking production schedules to environmental conditions, or correlating supplier data with equipment performance. A standardized framework helps ensure these relationships are captured systematically, creating a clear path for AI to extract meaningful insights.

Striking the right balance between too little and too much context is equally important. Insufficient detail can lead to oversimplification, while excessive complexity can overwhelm systems, obscure patterns, and reduce processing efficiency. By organizing relationships and metadata consistently, a consistent data model ensures that AI systems can process the right level of information proficiently, leading to better outcomes without unnecessary overhead.

The technical and organizational challenges of creating this connected model are significant but manageable with the right approach. Teams must define data relationships, integrate metadata across systems, and ensure compatibility with analytics platforms.

As manufacturing evolves into an AI-driven industry, contextualization becomes essential to unlocking its full potential by giving data meaning and providing AI systems the insight needed to operate strategically.

Contextualized Federated Data at the Edge and Beyond

When data federation, edge computing, and contextualization converge within a manufacturing ecosystem, they redefine how facilities operate and work together. As the architecture matures, it shifts from simply gathering and analyzing data to enabling a system that continuously learns, adapts, and improves across the organization.

This collaboration begins at the local level. Edge devices collect continuous streams of data directly from equipment and production zones, enabling real-time monitoring and decision making. These new devices are more than the passive sensors of the past. They're equipped with intelligence that processes data as it's generated, detecting patterns and anomalies instantly. Local teams bring an added layer of expertise, interpreting these patterns within the specific context of their operations. For example, they understand when a slight fluctuation in machine performance signals a potential issue or when it's inconsequential.

While edge systems excel at local optimization, manufacturing requires more than isolated insights. True operational excellence comes from shared learning across facilities. Each facility trains AI models using data unique to its equipment, processes, and environment, while sharing only the aggregated insights, not the raw data itself. These shared insights help other facilities learn and adapt without exposing restricted information or process specifics.

Take predictive maintenance as an example. One plant might use its historical data to train a model that accurately predicts failures under specific operating conditions. The insights derived from this model can then be shared across the organization. Another facility with similar equipment and operating conditions can use these learnings to optimize its own maintenance schedules, even without accessing the original data. This decentralized learning approach allows the organization to improve as a whole while respecting the uniqueness and privacy of each site.

This concept shines even brighter in quality management. Imagine discovering that subtle environmental changes, such as increased humidity during the summer, are linked to minor defects in a specific product. By sharing this finding through a federated network, other facilities with similar conditions can proactively adjust their operations, avoiding defects before they occur. There's no need for detailed data exchange, the shared insight is enough to drive actionable improvements.

As each facility solves problems and uncovers insights, these learnings contribute to a growing collective knowledge base. One plant might develop a more efficient way to manage material variability, while another fine-tunes its energy usage during peak production times. These discoveries flow through the network, enabling every facility to benefit from the organization's collective expertise.

Partnership across facilities also strengthens resilience. When one plant identifies issues with incoming materials, such as a supplier sending out a batch that doesn't meet specifications, it can alert others instantly. Each facility can respond based on its specific impact, but they all gain from the early warning. This interconnectedness ensures the organization is better prepared to handle shared challenges, minimizing disruptions across the board.

Bringing contextualization into the system adds valuable depth. When one facility shares a lesson learned, it comes with the full relevant picture, such as conditions, materials, and processes. A statement like "humidity affects quality" becomes far more useful when paired with the specifics behind it. This kind of clarity turns shared insights into actions that can be applied quickly to improve operations elsewhere. Patterns emerge and new opportunities take shape when these contextualized insights are aggregated and interpreted by centralized AI models.

Scaling this approach works best as an iterative process. The first step is ensuring each facility can effectively process its local data. From there, connecting facilities with similar processes or shared challenges allows for broader insights. As teams see the benefits of shared learning, confidence in the system grows, making it easier to expand across the enterprise.

People remain central to this transformation. Local experts provide the knowledge that makes insights actionable. They recognize which factors matter most in their operations and ensure knowledge sharing remains practical and relevant. Their expertise fuels the system with insights to shape its success. By valuing their input and demonstrating the impact of integrated data, organizations foster a culture of cooperation that drives continuous improvement.

Looking ahead, these connected systems will enable even greater possibilities. AI models will become more sophisticated, anticipating outcomes across entire operations rather than isolated systems. Predictive capabilities will extend beyond maintenance and quality control to include strategic decisions, such as resource allocation and production planning. Systems will dynamically adapt how knowledge is shared, tailoring insights to evolving production needs and shifting market conditions.

By building on a foundation of contextualized, federated data at the edge, manufacturers can create a scalable infrastructure that supports AI solutions across the enterprise.

Designing a Scalable and Context-Aware Architecture

Emerging AI solutions depend on a well structured foundation that brings together data from across the full manufacturing environment and makes it useful where and when it matters. A strong architectural framework connects technologies like edge processing, federated access, and contextualization into one cohesive system.

This framework needs to do more than move data. It must strike a careful balance between centralized control and local autonomy. That means allowing real-time decisions to happen at the edge while still ensuring that information flows easily across the broader enterprise. It also requires thoughtful design to support security, uptime, and adaptability. As systems evolve, so should the infrastructure that supports them.

At the heart of this structure is a clear division between operational systems and enterprise systems. On one side is the Operational Technology (OT) network, which manages real-time control of machinery and production processes. This layer includes the controllers, sensors, and operator interfaces that keep everything running. Performance here is measured in milliseconds. Any delay can affect product quality or interrupt the flow of production.

The Informational Technology (IT) network, by contrast, handles business systems and analytics. Here is where planning, reporting, and optimization tools reside. These systems depend on secure data, consistent connectivity, and access to cloud-based platforms and external services. While speed matters, the focus here is more about visibility and coordination across the organization.

To connect these two layers safely, a secure integration zone sits between them. This boundary area, known as a DMZ, allows data to move between networks without exposing critical systems. It acts as a buffer, collecting production data and passing it along without creating new security risks. Systems like Manufacturing Execution Systems (MES) typically operate in this zone, bridging the gap by gathering data from the floor and sharing it with business platforms for reporting and analysis.

In the past, centralized network models tried to push all data through a single hub. This approach created bottlenecks and made systems too rigid to handle the pace of modern manufacturing. Today's architectures take a more distributed approach. On the factory floor, mesh networking allows devices to share information directly, cutting down on delays. Software defined networking helps route traffic where it's needed most, while segmented networks keep sensitive systems secure without blocking necessary data.

This kind of flexibility makes the entire system more resilient and more responsive. It lays the groundwork for real-time data to move from the production line to the analytics platform and back again.

Once the network is in place, the next question is where to place computing resources. As discussed in previous sections, not all decisions need to come from a centralized system. Some need to happen instantly, right where the data is created. Others require a broader view across lines or sites. A layered computing approach is required to enable this.

At the site are computing servers that support the entire facility and integrate with edge devices. These machines collect and process data from many sources at once, pulling together insights from across the floor. A facility-level system might watch for bottlenecks, detect shifts in cycle time, or coordinate equipment usage across multiple lines. It sees the bigger picture from the site's perspective and helps operations stay efficient.

At the enterprise level, computing shifts from real-time response to deep analysis. These systems live in the cloud or in centralized data centers. They process historical records, run simulations, and train advanced algorithms that shape future decisions. Once these insights are developed, they feed back to the lower levels, helping edge systems work smarter based on what's already been learned.

Alongside this distribution of computing is a similar approach to storage. Just like processing, not all data needs to live in the same place or be stored forever. Data that flows quickly, like sensor readings, is stored close to where it's created. These edge storage systems use solid-state drives and other fast, compact hardware to hold high-speed data streams. They keep just enough information on hand to support live analytics and maintain short-term visibility. A machine might collect thousands of data points every minute, but only hold onto the detailed stream for a few days before summarizing and passing it along.

Facility level storage brings together these summaries and short-term records into a longer running view. This middle layer holds enough history to spot trends and run comparisons across shifts or batches. It supports process improvements and quality checks, giving teams the data they need to make informed adjustments without digging through outdated or irrelevant files.

As data flows upward through the architecture, storage evolves with it. Enterprise storage provides the long-term foundation for deeper insight. Whether hosted in cloud platforms or housed in corporate data centers, this layer holds historical records, aggregates cross-site comparisons, and supports broad reporting needs. It's where enterprise-wide patterns are uncovered and strategic planning takes shape.

Cloud solutions bring the advantage of scalability and easier access across locations. On-premises storage, on the other hand, might be chosen for its performance or tighter control over sensitive data. Both have their place, and together they support a balanced, resilient strategy.

This layered storage model ensures that each level plays its part. Operational data stays close to the process where it's needed most. Aggregated insights come together at the facility level. Enterprise systems retain historical depth and organizational reach.

With this constant movement of information between local equipment, on-site systems, and centralized platforms, security needs to be incorporated into the structure from the start. The best protection comes from a dispersed approach. Segmenting the network limits how far threats can spread. Encrypting data during transfer and while stored keeps it protected at all stages. Access controls ensure that only the right people can interact with sensitive systems. Regular patches and updates close the door on known vulnerabilities.

But protection doesn't stop there. Modern threats require smarter defenses. Behavioral analytics watch for unusual patterns that could signal trouble. Zero-trust frameworks remove assumptions about safety based on location and everything must be verified continuously. Secure-by-design principles help ensure that safety is built into every layer, not forced in after the fact.

Adaptability is just as important. New machines are constantly added, processes change, and new tools are incorporated. The architecture must be flexible enough to grow and shift without disruption.

At this point containerization and microservices are incorporated in the framework. Instead of building large, rigid applications, services are packaged in small, portable units. These can be updated, scaled, or deployed anywhere without rewriting code. It allows new functionality to roll out quickly and consistently, no matter where it's needed. For example, a predictive model might be trained in a centralized cloud environment, then packaged and pushed out to the shop floor. When improved, the update rolls out the same way with no downtime and no rework.

Designing a scalable, context-aware architecture requires a practical approach that builds value gradually while laying the foundation for growth. Initial integration efforts begin tying systems together, while early edge computing deployments enable real-time processing close to the action. Contextual layers begin to form as data is mapped to the Common Data Model. Information is linked to specific machines, products, and processes, creating the basis for meaning that both people and AI can interpret.

As momentum builds, new capabilities come online. Edge computing expands to cover more of the operation, delivering faster responses and reducing data bottlenecks. Contextualization deepens as local knowledge is embedded into the system, giving data the clarity it needs to drive intelligent actions. Advanced analytics begin to surface new insights, uncovering patterns that weren't visible before.

The most resilient systems don't stop evolving. As models mature, they're refined with real-world feedback, becoming more accurate and better aligned with the unique characteristics of each site. Proven solutions scale across the enterprise, bringing consistency and expanding benefits. New technologies are tested and added as needed, keeping the architecture modern and relevant. Throughout this process, progress is tracked against the original goals to ensure each step delivers measurable value.

By putting this architecture in place, manufacturers establish more than just a technology platform, they build a responsive framework that adapts with them. Each layer, from edge computing to contextual integration, works together to create a foundation where data flows securely, insights arrive faster, and decisions are grounded in real understanding. As more capabilities are added, whether it's knowledge systems, visual analytics, or AI optimization, the architecture supports continued progress without needing to be rebuilt.

Bringing It Together as an AI Enabler

With a flexible architecture in place, companies can use AI to create meaningful change across their factories. Data federation, local processing, and contextual awareness combine to help AI deliver both broad impact and site-specific precision. The scenarios in this section show how this integration already shapes day-to-day manufacturing.

One compelling example is improving product quality across an organization. When data from multiple facilities becomes federated, AI models can analyze patterns that traditional methods might miss. These models might uncover that a consequence of related factors such as slight variations in raw materials, specific environmental conditions, and subtle shifts in equipment calibration can produce the best results. This insight moves beyond theoretical analysis to real-world impact through localized intelligence.

At one plant, this could mean adjusting machine speeds to enhance assembly accuracy, while another facility optimizes cooling times based on the properties of the materials in use. Edge systems make these adjustments instantly, reflecting the local conditions at each site while staying aligned with broader enterprise goals. The outcome is an overall boost in quality, as every facility benefits from shared knowledge but applies it in a way that works for its unique setup.

This blend of global insights and local adaptability will reshape how factories manage their entire supply chain. There are many challenges in managing supply chains such as fluctuating demand, transportation delays, and material shortages. With AI processing data from suppliers, logistics partners, and internal systems, manufacturers can create a real-time model of this, helping to quickly adjust to any disruptions.

Local computing plays a key role in making these adjustments fast and effective. For example, if there's a delay in raw material deliveries due to weather conditions, tools running at production sites can make immediate decisions based on local needs. At a factory, the system might reroute materials from a different supplier, adjust the production schedule to optimize for available inventory, or even shift resources between production lines to maintain workflow. These local decisions ensure that the site can keep running smoothly without waiting for approval from a centralized system.

Analytics applications ensure that these decisions are made quickly and on the spot, based on local conditions. It takes into account real-time factors like inventory levels, equipment availability, and production priorities, allowing each facility to act independently without waiting for off-site direction.

This same combination of global insights and local adaptability can be applied to energy management. Instead of simply tracking energy consumption, AI evaluates patterns across facilities, factoring in fluctuating energy prices, weather conditions, and renewable energy availability. These insights inform strategies to optimize energy use across the operation. Locally, a factory might stagger equipment startup times to avoid peak pricing or align energy intensive processes with periods of renewable energy availability. This approach builds a sustainable, strategically aligned energy system.

The true strength of AI lies in its ability to connect the dots. Decisions to reduce energy costs, for example, are made with an understanding of the full impact on staffing, inventory, and supply chain performance. AI ensures these interconnected factors remain in balance, finding solutions that improve the overall operation rather than optimizing one area at the expense of another.

Collaborative robotics is another area where these capabilities shine. When data from robots across facilities is federated, AI can identify patterns that improve performance. One plant might discover a movement sequence that reduces wear on components, while another finds a way to enhance efficiency during specific tasks. These insights are shared across the organization and tailored locally by edge systems. Each robot adjusts its actions based on site-specific variables like material properties, factory layout, environmental conditions, or equipment configurations.

AI doesn't just optimize robot operations, it dynamically coordinates workflows between machines and people. For instance, during peak demand, robots might handle repetitive tasks, freeing human workers to focus on quality checks or troubleshooting. This collaborative model creates a production environment that is both adaptable and resilient, responding in real-time to changing demands.

Another area gaining momentum is adaptive workforce training. As factories introduce new technologies and processes, keeping operators up to speed becomes a constant challenge. AI can help by analyzing operational data to identify where skill gaps impact performance. For example, if production slows during certain shifts or errors rise with specific tasks, AI systems can trace patterns back to experience levels, training history, or onboarding progress. Localized systems can then deliver tailored guidance based on the task at hand and the operator's current proficiency. This may include step-by-step instructions, real-time prompts, or even AR-based walkthroughs. This targeted approach supports continuous learning on the floor, reducing errors while accelerating confidence and capability. Over time, the system adapts to each operator's pace, creating a personalized development path grounded in real production data.

AI-driven manufacturing moves the industry beyond the current operating model centered on reacting to problems as they arise. It enables factories to anticipate challenges, leverage opportunities, and create a dynamic system that thrives on continuous learning. By seamlessly integrating localized insight with enterprise-wide intelligence, manufacturers gain the agility to address immediate issues while laying the groundwork for continued growth.

Insights in Action

Norman wiped the sweat from his brow as he stood in front of the old stamping press, a reliable workhorse that had been on the factory floor longer than most of the employees. Today, though, it was acting up and producing parts that were just slightly out of spec.

"Come on, old girl," Norman muttered, patting the machine. "What's got you all riled up today?"

He'd been through the usual checks. The die seemed fine, the hydraulics were working correctly, and the material feed was consistent. Yet something was off, and he couldn't put his finger on it.

Just then, Rita walked by with a tablet in hand running the new troubleshooting app. "Hey Norman, want to try out our new system? It might help."

Norman eyed the tablet suspiciously but took it. "Alright, let's see what this new gadget will tell me."

As he input the details of the problem, the tablet's screen came to life. It pulled up more than the press's maintenance history and current sensor readings, but also data he hadn't considered. This included the day's temperature and humidity, the specific batch number of the raw materials, and even the subtle variations in the power supply over the last 24 hours.

Norman's eyes widened as he saw the connections forming on the screen. The combination of unusually high humidity, a slightly different composition in the latest material batch, and minor power fluctuations were all contributing to the press's inconsistent performance.

Norman shook his head in amazement. "It's not just one thing, it's everything together!"

Following the system's suggestions, Norman made a few adjustments to the press settings, taking into account the environmental conditions and material properties. He ran a test piece, and to his satisfaction, it came out perfectly within spec.

"You know," Norman said, handing the tablet back to Rita with a smile, "I've always said you need to listen to these old machines. Turns out, you just need the right translator. This new thing isn't replacing my experience, it's just helping me put all the pieces together faster."

As Rita walked away, Norman turned back to the press with a newfound appreciation. Maybe all those meetings about data and connectivity weren't just IT jargon after all!

Chapter 4

Creating a Unified Data Layer

||||||||||||||||||||||||||||||||||||||||||||||||||||||||||||||||||||||||||||||||||||||

Introduction

Imagine a manufacturing plant running at full steam late into the night. An unexpected vibration triggers an alert. Within seconds, the maintenance system pulls equipment history, production schedules adjust automatically, and quality parameters tighten on downstream processes. The night shift supervisor sees everything on a single screen. Not five systems, not three dashboards. One unified view showing exactly what's happening and what needs to happen next.

This is the power of a Unified Data Layer in action.

Most manufacturing operations today sit on goldmines of information they can't fully use. Sensor data streams continuously from equipment. Quality systems track every measurement. Supply chains generate continuous updates. Energy meters monitor consumption patterns. But when a critical decision needs to be made, finding the right information feels like archaeology. Digging through system after system, hoping the pieces fit together in time.

The UDL turns scattered signals into a coherent picture, where all manufacturing data converges and flows. Not just connected, but truly unified. Information arrives from every corner of the operation and becomes immediately accessible, contextual, and actionable.

When data moves through a well-designed UDL, it opens new possibilities. Patterns that were hidden in separate systems come into clear view. A small temperature change in one process shows its effect on quality further down the line. Minor variations in raw material properties link directly to shifts in production efficiency. Maintenance records highlight issues before they cause failures. The whole operation becomes clearer, more responsive, and smarter.

AI and advanced analytics flourish in this environment. Machine learning models no longer work with fragments of information. They access complete operational context, making predictions that actually match reality. Optimization algorithms balance multiple variables simultaneously because all the data they need exists in one place, updated continuously.

Change comes when people share the same unified view. Operators spend less time searching for information and more time solving problems. Engineers see issues in full context instead of as isolated symptoms. Managers base decisions on what is happening now rather than on last week's reports. When everyone works from the same real-time intelligence, the whole organization moves faster and makes better choices.

This chapter looks at how to design a system that supports lasting change. It explains how to build an architecture that can handle continuous streams of live data without strain and ensure information reaches the right place at the right time. The outcome is a new way of working where data powers repeatable best practices across the company.

Manufacturing is entering an era where success comes from how quickly and intelligently organizations can act on their data. The Unified Data Layer makes that speed and intelligence possible, turning information from a challenge to manage into the most powerful asset in modern manufacturing.

Foundations of the Unified Data Layer

With many of the foundations of the Unified Manufacturing Data Architecture in place, it's time to explore the component that ties it all together, the Unified Data Layer. This is the backbone of the architecture. It's what turns the individual capabilities we've discussed such as purpose-built Common Data Models, contextualization, data federation, and edge computing into an integrated data ecosystem.

As data consuming AI use cases grow, the challenge of supporting them shifts from managing individual CDMs to integrating them into a cohesive system. The UDL harmonizes across domain-standardized schemas defined by CDMs, preserves operational context, and exposes consistent access patterns to downstream consumers facilitating seamless data flow.

What sets this approach apart is its ability to unify data while respecting the unique needs of each domain. Rather than forcing all systems into the same mold, it connects them in a way that preserves their individual value and strengthens the greater whole.

Understanding the strength of the UDL starts with seeing how it works alongside the Common Data Models. Each serves a distinct purpose, and together they provide a scalable way to manage data across complex manufacturing operations. Common Data Models focus on standardizing information within specific domains. A Production CDM, for instance, ensures that data from different machines or sites follows the same units, field names, and validation rules. The same idea applies across quality, maintenance, and supply chain systems. Each domain structures its data to keep it consistent, clean, and ready to use.

The Unified Data Layer connects these standardized domains and ensures they can work together. If one data model refers to a production unit as a "batch" and another calls it a "lot," the UDL recognizes they mean the same thing. Importantly, the UDL doesn't redefine domain schemas, it maps and links them through shared reference data and lightweight translation layers when needed. This allows each team to continue using the terms and structures they're familiar with, while still aligning with the broader ecosystem.

This is possible through shared reference data such as central records for assets, materials, and processes that appear in more than one system. A mixer might be labeled "Machine AB-200" in production, but logged as "Line 1 Mixer" in maintenance. The UDL links these identifiers through a shared equipment master, ensuring data about the same machine can be correlated, regardless of where it came from.

This becomes especially useful when something goes wrong. Imagine a quality engineer investigating an increase in product defects. The Quality CDM has already standardized how those defects are categorized. Using the UDL, the engineer can access related data from other domains such as recent equipment changes in production, missed calibrations in maintenance, or shifts in supplier lots from procurement. Each of those systems is organized by its own data model, and the UDL ties them together in a way that makes the full picture clear.

None of these systems needed to conform to a single structure. The production team keeps its naming conventions. The maintenance group doesn't need to change how it logs service activity. The UDL works behind the scenes to connect these worlds, making the relationships visible without forcing uniformity.

This layered approach keeps things manageable as the organization grows. Domains stay in control of their data, while the UDL focuses on connection and context. For advanced analytics and machine learning, this structure is essential. It provides models with access to high-quality, harmonized data across domains, enabling deeper insights and more accurate predictions that reflect the full complexity of modern manufacturing.

Consider a manufacturing enterprise with multiple facilities, each using its own production data model tailored to local processes and equipment. While these site-specific models optimize local operations, they can create barriers when enterprise-wide insights are needed. The UDL bridges these gaps by aligning the different CDMs, allowing data to be shared and understood uniformly across the organization.

This integration of cross-domain data enables advanced analytics and AI applications to access comprehensive datasets, uncovering patterns and insights that would remain hidden within isolated datasets. For instance, correlating production data with supply chain information can lead to more accurate demand forecasting and inventory management.

Imagine a manufacturing operation as a bustling city. Each department, machine, and system is like a neighborhood, full of activity and purpose. But without roads, traffic signals, and bridges connecting them, the city becomes chaotic, with each part working in isolation. The Unified Data Layer is the city's infrastructure, creating the pathways that link everything together. It ensures the smooth flow of information, enabling collaboration and coordination across the enterprise.

To understand where the UDL fits within the broader architecture, let's take a closer look at it. The following diagram shows its role as the base of this unified approach, highlighting how it integrates diverse sources into a cohesive framework to support multiple data needs.

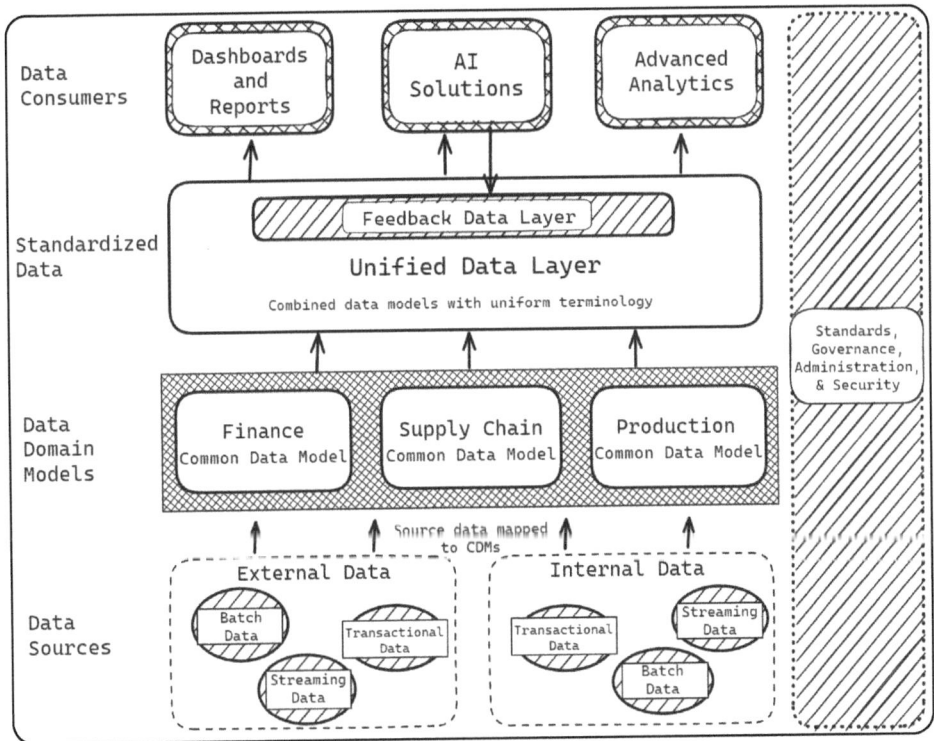

Figure 4-1 UDL Data Source to Data Consumer Flow

This diagram presents an overview of how the components within the

Unified Manufacturing Data Architecture work together. Each layer represents a crucial stage in this process, from initial data generation to its final use in responding to events and supporting strategic goals. Let's explore the components in more detail to understand their roles within the architecture.

At the base of the architecture are the data sources. These include internal systems like ERP, MES, SCADA, and LIMS, each generating unique types of data such as transactional records, batch updates, and real-time streams. External sources play an equally vital role, bringing in supplier data, logistics updates, and market forecasts. Together, these diverse inputs represent a broad view of the manufacturing ecosystem, illustrating the UMDA's capacity to integrate structured and unstructured data from a variety of origins.

Sitting above the raw data layer are the data domain models. These are standardized frameworks tailored to specific areas such as finance, supply chain, and production. These domain-specific Common Data Models ensure consistency by translating disparate data into a shared structure and language. Acting as maps, they organize the havoc of incoming information and prepare it for seamless integration into the Unified Data Layer.

The Unified Data Layer sits at the center of the architecture. It provides a unified access layer where harmonized views across standardized domain data are brought together, offering a complete view of the organization. Within this sits the Feedback Data Layer (FDL), a specialized component that captures real-time insights and historical trends. This combination helps the UDL both manage data and support advanced analytics and AI models. The FDL will be covered in detail later in this chapter.

At the top of the diagram are the data consumers, the tools and systems that turn data into insights. This layer includes dashboards and reports for executives and managers, AI-driven solutions for predictive and prescriptive analytics, and advanced tools used by data scientists to uncover hidden trends and correlations.

The type of output depends on the work being done. Production focuses on machine performance and quality checks. Finance looks at budgets and profitability. Supply chain tracks inventory and deliveries. R&D studies materials and prototypes. Each area speaks its own language, so the underlying models must reflect those differences while still tying everything back to a shared view.

This enables data to serve the priorities of each team.

The challenge lies in ensuring these specialized models operate independently yet remain connected to the larger framework. The main goal of the UDL is to enable cross-functional collaboration without forcing teams to compromise on the unique details critical to their work. Production data, for example, can flow seamlessly into financial analysis to link machine efficiency with cost savings. Supply chain insights can integrate with R&D data to identify material innovations that improve both performance and logistics.

By allowing functional areas to maintain their individuality while contributing to a larger framework, the architecture creates an organization-wide synergy. Teams can work autonomously where needed but tap into a shared pool of information to uncover patterns, address challenges, and pursue opportunities that span outside of their domain. It's a balance of independence and alignment that ensures the system serves all stakeholders, empowering them to act with greater precision and confidence in the data.

One of the UDL's most powerful features is its ability to integrate real-time and historical data seamlessly. High-speed streams from sensors, controllers, and other devices are combined with long-term records stored in enterprise platforms. This fusion enables advanced applications like predictive maintenance, where real-time equipment data is compared against historical patterns to identify potential failures before they happen. Similarly, live production data can feed into supply chain adjustments, ensuring resources are allocated efficiently as conditions change.

Real-time data processing takes this empowerment to another level. In manufacturing industries where timing and precision are non-negotiable, the ability to respond without delay can mean the difference between success and costly setbacks. Immediate insights enable proactive adjustments that optimize processes on the spot.

Adaptability is another important component of the UDL. Facilities are constantly evolving, with new technologies, workflows, and data sources being incorporated. This architecture is designed to grow with these changes, seamlessly aligning new elements without disrupting the processes already in place. Its flexibility ensures that as operations expand, the data framework continues to deliver reliable value.

By seamlessly integrating diverse data sources, standardizing access patterns, and enabling collaboration across departments, the Unified Data Layer empowers manufacturers to harness the full value of their data. Whether improving legacy systems or supporting advanced analytics and AI, this connected approach enables continuous improvement.

The Four Pillars of Data Management in the UDL

To understand the role of the Unified Data Layer in turning information into usable knowledge, it helps to look at its main building blocks: ingestion, storage, processing, and access. These pillars define how the UDL works by tackling the biggest challenges in handling different types of data.

While our previous chapter explored data federation as a way to connect to data where it resides, the UDL takes a hybrid approach that combines federation with selective data movement. The four pillars we'll explore encompass both federated access and physical data integration depending on specific needs. This balanced strategy allows the UDL to leverage data virtualization where appropriate while also merging data when necessary for performance, analytics, or AI applications.

The first pillar, ingestion, ensures valuable data from every corner of the company can be incorporated into the system. This doesn't necessarily mean physically moving all data. In many cases, ingestion involves capturing metadata and establishing connections to data that remains in source systems. For critical or frequently accessed information, physical ingestion brings the data directly into the UDL storage. For other data, especially large volumes or information that needs to remain in source systems, federation provides virtual access without duplication.

The sheer variety and volume of data present unique challenges to large AI models. Real-time streams from IoT sensors monitor everything from machine vibrations to environmental conditions, while PLCs provide detailed metrics on production performance. Quality systems contribute data on defect rates and product specifications, and maintenance teams add layers of insight with their logs and observations. Each of these sources tells a piece of the story, and ingestion ensures that none of it gets lost or disconnected from its broader context.

This process reaches beyond the production floor. Enterprise platforms add data on ideal inventory levels, order progress, and resource planning. Supply chain systems contribute insight into delivery timelines and vendor performance. Even unstructured sources such as customer feedback, market indicators, and weather forecasts can be pulled in to provide valuable external signals.

To manage this range of inputs, the system aligns incoming data with the appropriate Common Data Model. As information arrives, it's matched to a domain-specific structure where production metrics follow the Production CDM, quality outcomes map to the Quality CDM, and Inventory data fits into its own model. These models don't operate in isolation. Cross-domain links connect them, making it possible to trace relationships between production processes, quality results, and material flows. This alignment provides a consistent framework that keeps data usable and connected, no matter where it comes from.

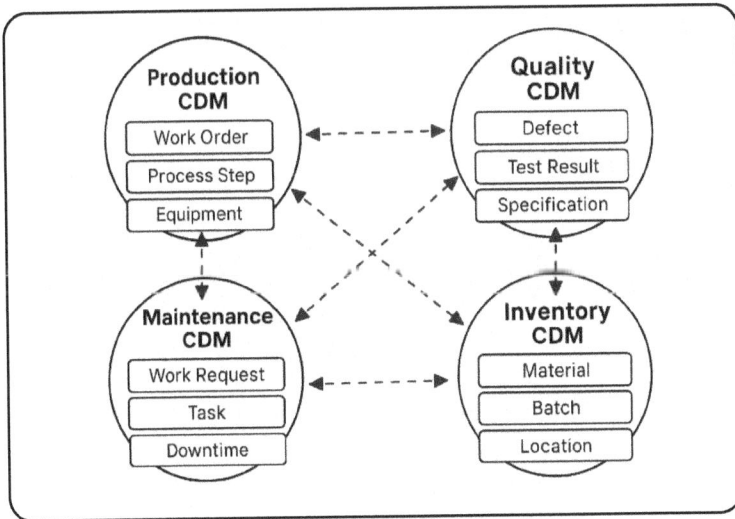

Figure 4-2 Cross-Domain CDM Connections

As the data is ingested, the system enriches each data point with metadata. Every record is tagged with identifiers like source, timestamp, and domain relevance. These tags preserve the context needed to connect events, trace patterns, and support analytics.

To handle these complexities, the architecture employs a combination of batch and real-time processing methods. Batch pipelines, like Extract, Transform, Load (ETL) workflows, efficiently manage large volumes of historical data by aligning it with the CDM frameworks during transformation. In contrast, real-time stream processing technologies manage high frequency inputs from sensors and transactional systems, applying the CDM mappings on the fly to harmonize diverse data streams.

By anchoring the ingestion process in the standardized data models, the architecture achieves both flexibility and consistency. This foundational step ensures that all incoming data, regardless of its origin or format, can be seamlessly integrated and prepared for meaningful analysis.

Once data has been ingested, the next consideration is where and how it should be stored, or whether it should be stored at all. Some information is brought directly into the Unified Data Layer for high-speed access, while other data is left in its original location and accessed through federation. The system must support both methods, choosing the best approach based on how the data will be used.

For example, high-frequency signals from vibration sensors might be stored at the edge in local databases to support real-time monitoring and fast response. Production records, which are used less frequently, might be stored centrally in an enterprise data lake or cloud platform. Data from supplier systems or regulatory archives might not be stored within the UDL at all, but accessed through federation when needed for analysis or reporting.

The hybrid storage model balances the capabilities of on-premises and cloud systems. Local edge storage handles sensitive and time-critical data like equipment performance metrics and real-time sensor inputs. This setup ensures rapid access, low latency, and tight security for immediate decision making and operational responsiveness.

Cloud storage complements these on-premises systems by offering a scalable solution for retaining large volumes of historical data and supporting enterprise-wide analytics. Together, this approach combines the immediacy of local processing with the scalability of centralized resources, creating a robust and flexible foundation for managing manufacturing data.

What sets this approach apart from traditional architectures is its ability to unify these diverse storage tiers into a seamless logical framework. Using advanced tools like data virtualization and metadata management, the system hides the complexities of physical storage, offering users a unified view of the data. For AI applications this unified perspective is invaluable since it allows data from different domains to be analyzed together, while preserving the context established by domain-specific models.

The third pillar, processing, converts raw data into actionable insights. Here, processing combines real-time data streams with historical trends to support advanced analytics and AI. For example, predictive maintenance relies on processing capabilities to identify subtle patterns in equipment performance. The system interprets potential issues within the broader operational context, suggesting precise interventions.

Consider a factory needing to balance production scheduling with workforce availability. The processing system analyzes current data from production lines such as output rates, equipment utilization, and quality metrics, alongside workforce inputs like shift schedules, skill sets, and attendance. By integrating this information, it identifies where production bottlenecks are forming and which skilled operators are available to address them.

For instance, if a production line is experiencing slower than expected output due to material handling delays, the system might suggest reallocating team members with forklift expertise to the affected area. It simultaneously evaluates downstream impacts such as whether quality checks or packaging will need extra staffing to maintain flow.

This processing capability allows manufacturers to adjust workforce deployment dynamically, ensuring smooth operations without overburdening employees. The result is a responsive, efficient production environment where human and machine resources are used effectively, reducing delays and maintaining output quality.

Processing also drives adaptive raw material management. For instance, real-time production data might highlight a sudden shift in demand, prompting the system to adjust supply chain logistics on the fly. Materials can be rerouted to high-priority locations, production schedules can be modified to meet deadlines, and inventory levels can be balanced. This dynamic capability ensures uninterrupted operations even in the face of unexpected changes.

Finally, access is what turns all of this into tangible results. Data is only as valuable as the insights it delivers, and access ensures that the right people, systems, or AI models receive it in a form that can drive action. From real-time dashboards for operators to deep-dive analytics for data scientists, access tailors information to each user's needs.

Real-time dashboards give frontline teams the tools they need to monitor production and make adjustments as needed. Analysts and data scientists gain access to advanced platforms, enabling them to uncover trends and relationships across domains. For managers, the insights delivered provide a strategic view of operations, helping guide high-level decisions. AI models also benefit from this unified access, using contextualized data to identify patterns, predict outcomes, and recommend precise actions. Developers play a key role as well, leveraging APIs to embed these insights into custom applications, seamlessly integrating data-driven decisions into daily workflows.

This pillar also enables seamless machine-to-machine communication. Automated systems access shared data to respond dynamically to changing conditions, creating a manufacturing environment that is both adaptive and intelligent. This fosters a responsive manufacturing ecosystem, where processes can adjust dynamically to optimize performance and prevent disruptions.

Flexibility and scalability are central to this access framework. The system doesn't require consolidating all data into a single repository. Instead, it uses a virtual access layer that seamlessly integrates data from multiple locations. This ensures that users can query, analyze, and visualize information without worrying about its physical location or format. Whether data resides in edge devices, on-premises servers, or cloud platforms, it's available in a unified and intuitive interface.

When data is accessed through the UDL, the full context of that information comes with it. Every data point brings along the metadata that shows where it came from, what it relates to, and why it matters. So when a production manager reviews inventory levels, they can see those numbers in light of real-time demand forecasts and delivery schedules. This kind of holistic view helps each team make smart decisions while staying in sync with the rest of the operation.

While domain experts see data through their specialized lens, the UDL facilitates shared understanding by connecting perspectives. This interconnected approach fosters alignment, ensuring that departments can work together to achieve organizational goals.

UDL Architecture Planning

The path to implementing a successful Unified Data Layer begins long before the first system is connected. Thoughtful architecture planning serves as the blueprint that shapes how data will flow and deliver value across manufacturing operations. Much like designing a complex manufacturing facility requires careful consideration of material flows, equipment placement, and worker movement, planning a UDL demands strategic decisions concerning data pathways, processing locations, and access patterns. This planning phase establishes the critical foundation upon which all subsequent implementation decisions will rest, directly influencing how effectively the organization can leverage its data assets.

Begin by understanding the current data environment. Map out the flow of information within each domain such as Production, Supply Chain, Finance, and Maintenance. How is data stored? Where is it processed? What systems are currently in use? This assessment reveals inefficiencies and gaps that hinder collaboration and decision making. This detailed view of the current landscape will identify opportunities to improve workflows, standardize processes, and integrate disparate systems into a unified approach.

The next step is to prioritize key data sources. Not all information carries the same value, so the focus should be on the data that delivers the greatest impact. Domain experts can help identify what matters most. Real-time machine data may be essential for production, while financial metrics and inventory levels support planning and forecasting. External sources like market trends or customer feedback can also provide valuable insights that shape strategic decisions and long-term planning.

Once a clear picture of the current landscape and priorities has been captured, define objectives for the data layer. Each domain will have its own needs. Production teams may aim for better real-time visibility, while finance teams might seek integrated datasets for accurate forecasting. Align these goals with broader organizational priorities to create an architecture that supports both domain specific tasks and enterprise-wide initiatives.

Flexibility plays a critical role in maintaining alignment across systems. A modular architecture enables the organization to evolve without disrupting core operations. Approaches such as microservices and API-driven frameworks make it possible to upgrade components, integrate new technologies, or adjust to changing business requirements with minimal impact. This adaptability ensures that individual teams can use the tools best suited to their needs while still contributing to a unified, enterprise-wide data platform.

Equally critical is to consider plans for scalability from the start. As operations grow and data volumes increase, the architecture must remain capable of managing this expansion without performance bottlenecks. This could involve using elastic cloud storage for flexibility or incorporating distributed processing to handle high frequency data streams efficiently. A scalable system ensures the data layer stays robust and responsive, no matter how frequently the organization evolves.

Creating a secure data environment requires striking a balance between accessibility and protection. Sensitive information must be safeguarded without hindering collaboration across teams. Role-based access controls ensure that data is available to those who need it while restricting unauthorized access. Simultaneously, secure sharing mechanisms maintain the integrity of information as it moves between users and systems. This approach fosters a system that is both open and secure.

Establishing this secure foundation is essential before introducing more advanced capabilities. Once the right safeguards are in place, the architecture can begin to support a wide range of AI applications. Each of these tools brings its own set of data requirements, which must be considered early in the design to ensure the system delivers value where it matters most.

For instance, predictive maintenance tools need both live sensor readings and past equipment data, which may cover several years of records. Quality prediction systems must link process settings with inspection outcomes over numerous production cycles. Supply chain optimization relies on combining internal production details with external supplier and logistics information.

To meet these diverse requirements, data should be organized strategically. Real-time machine learning models, such as those forecasting equipment issues, benefit from time-series data stores. These can hold large volumes of historical sensor information while keeping the detail necessary to spot early warning signs. On the other hand, AI tools aimed at improving operations may perform better with data warehouses. Here, structured production and inventory data help in making accurate demand forecasts.

These considerations influence implementation choices. When setting up data ingestion for the Unified Data Layer, it's a best practice to focus on collecting high-frequency sensor data from critical equipment first. This data is vital for effective predictive maintenance. Regarding storage, using time-based partitioning can be helpful. Recent data, crucial for real-time AI, should be easily accessible, while older data can be archived cost-effectively but still available for model training.

The processing component also plays a key role in supporting AI. Unlike traditional analytics that depend on summarized data, AI models frequently look at detailed data closer to the source to find subtle patterns. The architecture should keep this degree of detail and support the processes that convert raw data into formats suitable for AI. For example, raw vibration data might be stored but also processed to extract features that make pattern recognition more efficient for machine learning.

Additionally, the architecture should support the entire AI lifecycle from design to deployment and ongoing monitoring. Data scientists need environments to create and test models using historical data. Operational AI systems require dependable, quick access to current data streams. The system should also include feedback mechanisms to track model performance, allowing continuous improvement as manufacturing conditions change.

Emerging paradigms like the Data Mesh enhance this architectural approach by shifting the focus from merely integrating and standardizing data to actively managing it as a product. This concept re-imagines data not as a passive resource but as an actively maintained, high value asset. Under this model, each domain becomes the steward of its own data products, taking ownership of their quality, usability, and lifecycle.

For example, a production team might develop a comprehensive "Production Efficiency Data Product" that includes machine performance metrics, downtime trends, and real-time production rates. This product would come with clear documentation, interfaces for seamless access, and service-level agreements that ensure reliability for its consumers, whether those are internal teams, external partners, or automated systems.

Treating data as a product fosters a culture of accountability and precision. Domains think of their data as something to be maintained, improved, and delivered with the end user in mind. This approach also supports versioning, allowing teams to track updates in collection methods or data structures, ensuring transparency and continuity over time.

Incorporating Data Mesh principles builds on the architecture we've established rather than replacing it. The existing components already provide the foundational tools for storage, processing, and structured delivery that distributed data products require. Data Mesh concepts can enhance this foundation with practical additions such as a centralized catalog where users can discover available data products along with their service levels and usage metrics, self-service interfaces that enable teams to access and use data independently without IT support, and performance monitoring that tracks the reliability and value of each data product across the enterprise.

Together, these elements encourage stronger data ownership and a more user-friendly approach to data management. Domains take responsibility for curating their own data products, which feed the analytics and AI systems.

The success of this system lies in its ability to bring together diverse tools and technologies under a shared framework. While different domains may use varying platforms the architecture ensures consistency in governance, quality standards, and semantic alignment. This allows for seamless data integration and analysis, no matter where or how the data is generated.

Integrating Data Sources into the UDL

Building the Unified Data Layer is a significant achievement on its own, but its value comes from connecting it to the broader manufacturing ecosystem. This involves creating a seamless flow of data between systems, teams, and technologies. The process of integration is where fragmented domain-specific data aligns with a cohesive framework.

It's not enough to simply aggregate data, it must be harmonized, contextualized, and enriched to ensure usability across domains. For example, aligning data from production with quality metrics or linking supply chain disruptions to real-time inventory levels requires careful mapping and coordination.

Predictive models, anomaly detection systems, and prescriptive analytics rely on integration to provide the breadth and depth of information necessary for accurate results. A unified approach to integration ensures that AI has access to consistent, context-rich datasets.

To better understand how this integration is achieved, the following diagram provides a high-level overview of an example data flow within a Unified Data Layer architecture:

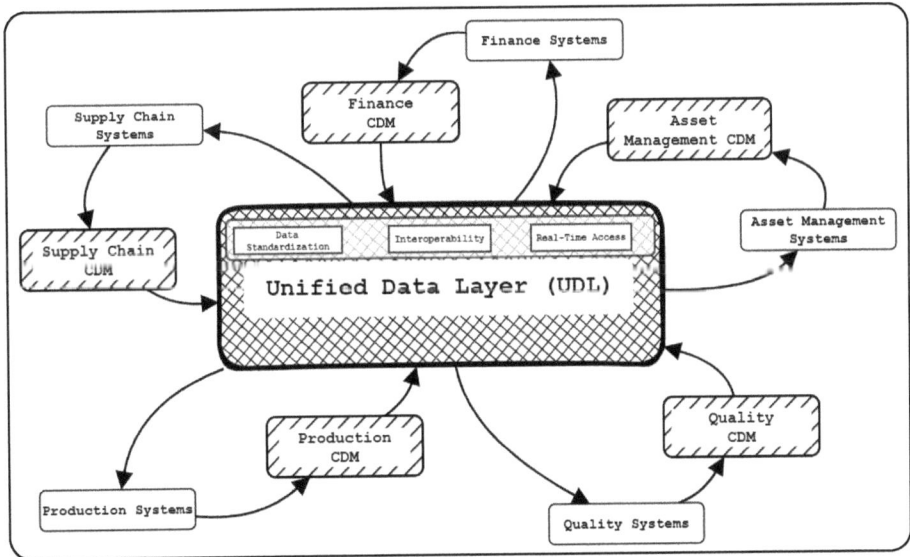

Figure 4-3 CDM to UDL Data Cycles

The diagram illustrates how key business systems integrate within the Unified Data Layer using Common Data Models. At the core of this architecture, the UDL functions as a logical hub, providing an authoritative view across federated and ingested sources. Surrounding the UDL are systems representing critical domains such as Finance, Supply Chain, Production, Quality, and Asset Management. Each of these systems feed into the UDL through the relevant CDM, ensuring a seamless flow of information across the enterprise.

Common Data Models are essential to this architecture. Acting as intermediaries, they harmonize data from various systems into a domain-consistent structure defined by the domain-centric data models, which the UDL connects using shared enterprise reference data (e.g., equipment, materials). This standardization ensures interoperability, enabling insights that span multiple domains.

The diagram emphasizes the dynamic, two-way nature of this integration. Inbound data flows from source systems into the UDL, passing through the appropriate CDM for transformation and alignment. For example, data such as equipment performance metrics or production cycle times from a Manufacturing Execution System (MES) is processed through the Production CDM, ensuring its fields are mapped to enterprise reference keys and relationships for cross-domain joins. This process guarantees that all incoming data, regardless of its origin, maintains a consistent format and contextual meaning.

Outbound flows carry standardized and enriched data from the UDL back to connected systems. Financial teams might use operationally enriched cost data for forecasting, while supply chain managers rely on live inventory data contextualized with market demand projections. This bi-directional exchange ensures that every function within the organization has timely and accurate information tailored to its needs.

The path to achieving this begins with a clear map of the organization's data landscape. This means identifying every data source, understanding what makes each one unique, and defining how they connect to core operations. High-speed inputs like IoT signals from the production floor need stream processing to manage continuous data flows in real-time. Meanwhile, batch processing suits data that arrives in intervals, such as quarterly financial summaries or historical performance records. Some types, like transactional updates or event driven changes, call for a mix of both. For example, Change Data Capture (CDC) can track live updates while combining stream and batch processing brings that data into the Unified Data Layer smoothly and efficiently.

As mentioned earlier in the chapter, integration within the UDL doesn't always require centralizing data in a single repository. In many cases, a data virtualization layer creates a seamless access point for distributed datasets. This method is especially valuable when data must remain within specific systems due to regulatory or operational requirements. Virtualization delivers the benefits of integration while preserving the flexibility of distributed storage, ensuring all data stays accessible for analysis.

Technically, the UDL achieves integration through a robust combination of tools and technologies, including Extract, Transform, Load (ETL) process-es, Application Programming Interfaces (APIs), and microservices architecture. Each of these components contributes to building a framework that is both flexible and resilient.

ETL processes play a vital role in automating data movement and trans-formation. These workflows are tailored to the specific needs of each domain. For example, the production domain may use high-performance ETL tools to process sensor data from MES systems, enabling real-time insights. Finance teams might employ structured ETL workflows designed to integrate and vali-date accounting records, ensuring consistency and accuracy.

APIs serve as the communication highway, enabling real-time data sharing between systems. Each domain can expose APIs designed for its unique data requirements. A quality management system might provide APIs to share de-fect rates or inspection results, while supply chain systems expose inventory levels and supplier performance metrics. By orchestrating these APIs, the UDL ensures seamless interactions across domains, enhancing both efficiency and interoperability.

Microservices architecture adds agility and scalability to the integration process. By breaking down complex integration tasks into smaller, independent services, this approach allows for targeted development and deployment. For example, mapping product codes to enterprise master data across domains, while another aggregates and analyzes production metrics in real-time. Each service operates independently, enabling continuous updates and scaling without disrupting the broader system.

The flexibility of this integration approach shines in scenarios like on-boarding a newly acquired manufacturing plant. Existing systems at the plant can continue running without interruption while gradually connecting to the unified data ecosystem. By leveraging APIs or custom microservices, the plant can align with the enterprise-wide architecture at a manageable pace. This phased approach minimizes disruptions, allowing operations to proceed smoothly while laying a foundation for AI-driven innovation.

Greenfield sites, on the other hand, present an opportunity to implement new technologies from the start. Free from legacy constraints, these sites can deploy advanced integration methods such as real-time analytics tools, data mesh principles, and comprehensive microservices frameworks. This strategy enables them to fully align with the enterprise's data strategy while tailoring modern solutions to meet specific operational needs.

The combination of ETL workflows, APIs, and microservices forms a dynamic integration framework that balances adaptability with speed. Each domain customizes its integration processes to meet its specific needs while contributing to a cohesive enterprise-wide ecosystem. This tailored approach ensures that data flows seamlessly across the organization without compromising the unique requirements of individual systems.

As the interactions from these systems converge, maintaining data quality and governance becomes a key to successful integration. Processes must be implemented to resolve inconsistencies, fill in missing values, and correct errors in source data before it enters the unified framework. Assigning clear ownership and stewardship roles within each domain creates accountability, ensuring that data remains accurate, relevant, and ready for use.

Prioritizing integration efforts based on business value ensures meaningful results early on, building momentum for long-term success. Start by focusing on data sources that directly impact critical processes or offer significant opportunities for cross-domain insights.

For instance, linking production and quality data can deliver early wins through real-time monitoring and process refinement. Building on that foundation, organizations can gradually bring in supply chain metrics, asset management data, and financial records. This incremental rollout reduces disruption while steadily expanding the usefulness of the entire data ecosystem.

The circular exchange between the UDL and each domain-specific Common Data Model reinforces one of the architecture's core strengths of making data accessible, standardized, and easy to share across groups. Insights generated in one area can be quickly applied in others, improving outcomes throughout the organization.

Capturing and Applying Learnings with the Feedback Data Layer

Understanding and continuous improvement are key elements of productive manufacturing. Every minor adjustment to a process and every piece of insight gathered presents an opportunity to refine processes. But without a structured way to capture and act on those lessons, valuable knowledge can slip through the cracks. What's needed is a smart memory system that collects these insights, preserves them, and makes them actionable.

That's the role of the Feedback Data Layer (FDL). It serves as the responsive layer that gathers input from across the operation and channels it back into the system. As part of the Unified Manufacturing Data Architecture, it builds on the foundation of the UDL, enabling responsive processes that learn and improve over time.

While the Unified Data Layer handles raw and standardized information, the FDL focuses on actionable intelligence. It becomes the home for insights generated by AI, such as recommendations for predictive maintenance or adjustments to production parameters. But it doesn't end at automated data learnings. It also gathers human input such as operator observations, manual overrides, and notes on anomalies. This blend of machine intelligence and human expertise creates a comprehensive perspective of the manufacturing environment.

Every interaction with this layer adds to the collective intelligence. When a smart system suggests a change that improves efficiency or quality, the result is recorded, creating a reference point for future decisions. If an adjustment doesn't deliver the desired outcome, that lesson is logged too, helping refine future recommendations. This cycle of learning ensures that each insight, whether a success or not, helps the system evolve.

Importantly, the FDL keeps original data intact. Raw inputs remain clean and unchanged, while the system builds a growing library of insights and outcomes. Teams can trace each recommendation back to its source, evaluate its impact, and understand its rationale. This traceability creates a transparent, living record that improves over time, offering teams a foundation for trust and informed decision making.

The Feedback Data Layer plays a critical role in making intelligent systems smarter over time by combining contextual data with associated results and looping them back into the data ecosystem so that AI models can continuously improve. This layer doesn't rely on new infrastructure. Most organizations can build it using existing platforms, as long as they can support flexible data formats and event driven flows.

Many different types of feedback need to be handled in the architecture. Machine data, performance logs, and sensor readings arrive in structured formats. Operator input, maintenance notes, or contextual observations tend to be unstructured. A well-designed feedback layer organizes everything into a consistent stream of events, each tagged with what happened, when it occurred, who or what reported it, and what followed.

Many early adapters start with a focused high impact use case, like a bottleneck in production or a quality issue that AI can help resolve. These early wins provide useful patterns that strengthen models and improve data analysis. To scale, the system should support real-time input with minimal friction. That might mean lightweight interfaces, like guided forms, embedded into existing dashboards or mobile tools.

Integration points for this loop of insights may already exist within system environments. The key is to make sure this knowledge is routed to where it can make a difference. This may be a predictive maintenance model, a control system, or a team reviewing weekly performance. As more feedback flows through the system, visualization tools become essential. They help highlight patterns that AI can act on, and they provide clarity across multiple sites or processes.

Partnership lies at the heart of the Feedback Data Layer. It elevates insights into shared knowledge that benefits the entire organization. When a production bottleneck is identified, the system ensures that every impacted team receives the information they need. Production teams can take immediate action to address the issue, while other departments use the same data to adjust their workflows.

Human input plays a critical role in the success of the Feedback Data Layer. Consider an example where experienced operators see small issues that data alone can't reveal. When they make manual adjustments to resolve unusual situations, the system captures those actions and the context behind them. Over time, automated processes learn from these interventions, making their future suggestions more practical and relevant.

The FDL excels at understanding how adjustments in one area impact others. For example, a change in energy use on one production line could subtly affect material flow or product consistency in downstream processes. By capturing and analyzing these interdependencies, the system identifies innovations that can be expanded across the organization. If one site discovers an efficient way to reduce waste, the feedback system helps translate that success into actionable strategies for other locations, tailored to their unique conditions.

This layer seamlessly connects quick local decisions with larger, long-term improvements. Frontline teams can respond immediately to live data, addressing operational needs in real-time. The system aggregates insights from multiple facilities, building a knowledge base that shapes company-wide strategies. Every change is recorded with a clear explanation of what was done and why. This transparency is invaluable for audits, performance reviews, and continuous improvement efforts. Teams can make adjustments with confidence, knowing they can trace outcomes back to specific actions and refine their strategies accordingly.

As the platform expands, so does the power of this system. Each facility contributes unique lessons, adding depth and nuance to the shared pool of insights. Smart systems refine their ability to detect patterns as they encounter new situations, making their recommendations more targeted and impactful. This feedback process becomes more focused with more data, delivering actionable insights tailored to the needs of specific teams or locations.

This collaborative growth extends well beyond daily operations. When organizations adopt new technologies or introduce new products, the FDL helps capture valuable lessons that can be shared across the enterprise. Teams developing innovative methods can pass along their findings in a way that protects sensitive details, allowing the entire organization to progress together.

The longer this system operates, the stronger the connection between people and technology becomes. By capturing both automated suggestions and human insights, the system creates a cycle of mutual trust and continuous learning. Teams trust the technology because they see their expertise reflected in its recommendations, while the system evolves to better align with the realities of manufacturing. This partnership ensures that every improvement is practical in the real-world context of the manufacturing site.

Building this kind of system requires time and careful planning, but the rewards are far-reaching. Insights no longer stay locked within individual teams or locations. They flow freely to where they can have the most impact. Each success adds momentum to the cycle of improvement, speeding up innovation and helping the entire organization work smarter and more efficiently.

Looking to the future, the value of the Feedback Data Layer will only grow. Every decision, adjustment, and insight will add to the collective intelligence. By capturing and sharing these lessons effectively, we're creating an environment where daily activities can be leveraged for lasting improvements. This feedback foundation enables the AI collaboration we'll explore next, where human expertise and machine intelligence work together effectively.

AI as a Collaborative Partner

Something exciting happens when artificial intelligence works together with the Unified Data Layer and the feedback infrastructure. This combination creates a living system that grows smarter every day. AI shifts from being a basic analysis tool to becoming an active partner in improving data usability. It continually looks for better ways to connect information, spot patterns, and generate insights. AI helps build a system that remembers what works, learns from experience, and suggests smarter ways to operate. This creates a cycle of continuous improvement that makes the entire manufacturing operation more intelligent as it matures.

Continuous adaptation sits at the center of this collaboration. The Unified Data Layer gives AI a consistent foundation of structured information standardized by the Common Data Models and harmonized across domains. With that clarity, AI can spot useful patterns, find gaps, and even suggest ways to improve how data is organized. The Feedback Data Layer captures what happens next, how people respond to those suggestions, whether adjustments worked, and what results they delivered. When a new data source comes online, for example, AI can evaluate it and recommend where it fits within the existing models or whether new extensions are needed. As feedback flows in, those recommendations get better, helping the system keep pace with changing business needs.

Beyond structural refinement, AI improves data discovery and integration. For example, when incorporating a newly acquired facility's systems, AI analyzes the data structures and identifies connections to existing models within the UDL. If a Manufacturing Execution System (MES) includes a "cycle_time" field, AI might link it to "production_duration" in the current production data model rather than storing it as a new value with overlapping meaning.

Machine learning helps make sense of the complexity buried in manufacturing data. Classification models spot recurring patterns across different data sources and suggest how they should connect to existing models. Clustering techniques reveal natural groupings that help refine domain structures. Deep learning, especially tools originally built for language processing, can recognize when two fields mean the same thing, even if they're labeled differently. To keep things running smoothly, anomaly detection algorithms scan for unexpected patterns that could signal data errors or system faults.

Equally important is the collaboration between AI and human expertise. As they work together, they create a feedback loop that combines automated computational speed with the insights of those familiar with the equipment and processes. While AI processes vast datasets and detects patterns, human oversight validates these findings, ensuring alignment with real-world constraints. Over time, this collaboration enhances the reliability of insights and strengthens confidence in the broader data architecture.

By automating repetitive tasks, AI allows domain experts to focus on high value priorities. When the tool proposes mappings, adjustments, or corrections, a person ensures these recommendations are relevant and operationally sound. For instance, a process engineer might adjust an AI suggested mapping to reflect specific workflow details or reject it if it doesn't meet targeted goals. This interaction between AI's computational power and human judgment creates a high integrity system capable of adapting to the complexities of real-world manufacturing.

This collaboration fosters continuous improvement. Each validated adjustment feeds back into the AI, enhancing its ability to make more accurate and relevant suggestions in the future. Over time, this iterative process builds a sophisticated and adaptive data ecosystem. For example, AI might analyze historical production data to identify conditions that consistently yield high-quality results. These benchmarks can then guide future operations. Similarly, when defects arise, AI examines contributing factors, uncovers risks, and proposes corrective actions, ensuring the lessons learned inform future strategies.

Even failures become opportunities for growth. Anomalies in equipment data, for example, might signal early signs of wear or misalignment. AI detects these patterns and provides actionable recommendations to mitigate potential disruptions, ensuring product quality remains intact. By learning from both successes and setbacks, AI refines its predictive and prescriptive capabilities, continuously improving its contributions to factory operations.

A vital element in this partnership is Explainable AI (XAI). Understanding the reasoning behind recommendations is essential for any user that relies on the output of AI. XAI addresses this need by offering transparent insights that explain the data, logic, and reasoning that led to the output. This clarity empowers teams to act on AI-driven suggestions with greater confidence, knowing the underlying rationale aligns with their goals.

Consider a scenario where AI identifies an opportunity to improve a key process. Instead of presenting a stand-alone suggestion, XAI provides context: "The current temperature exceeds the optimal range by 5%, as defined in the Production data model. Additionally, defect rates have increased by 15% under similar conditions in the past quarter. Adjusting the feed rate downward by 7% should restore performance quality and adhere to the shift production schedule." This builds trust, enabling operators to act with confidence while reinforcing the credibility of AI-driven recommendations.

To illustrate this collaborative partnership in action, let's explore a theoretical example of how AI might work within a Unified Data Layer in a manufacturing context. Imagine a facility that produces precision components where quality consistency remains a constant challenge despite standardized procedures.

In such a scenario, the UDL might integrate multiple data sources such as machine outputs, environmental conditions, material specifications, and quality measurements. During the ingestion phase, AI could analyze new data sources and suggest appropriate mappings to existing Common Data Models. For instance, when connecting a quality inspection system, AI might recognize that fields like "surface measurements" align with "finish parameters" in the established Quality CDM, streamlining integration without extensive manual mapping.

Within the storage layer, AI algorithms could continuously examine data relationships, potentially discovering non-obvious correlations between manufacturing conditions and outcomes. The system might identify that components processed during certain environmental conditions exhibit subtle quality variations. This pattern would easily be missed in manual analysis since it involves multiple interacting variables rather than simple cause-and-effect relationships.

The processing capabilities of the UDL would enable AI to develop predictive models based on these complex interactions. When conditions approach patterns associated with quality issues, the system could generate contextualized recommendations: "Current environmental readings combined with the material properties of the active production batch suggest increased risk of finish variations. Consider adjusting processing parameters or scheduling this production when conditions improve."

Through the access layer, these insights would reach production supervisors on their dashboards. Here the human-AI collaboration would become evident. A supervisor might review the recommendation and apply additional expertise along with the final decision as to how to proceed.

If implemented, the outcome of this decision would flow back into the system, allowing the AI to incorporate this human insight into future recommendations. Over time, this feedback loop could create a progressively smarter system where both quality issues and their solutions become more predictable.

Incorporating human insights into AI systems through the FDL enhances their predictive accuracy while fostering a more intuitive and trustworthy decision making process. This integration is particularly vital in sectors governed by stringent compliance standards, where the ability to trace and justify AI-driven decisions is imperative. Explainable AI addresses this necessity by providing clear, understandable rationales for each recommendation, ensuring that AI operations adhere to regulatory requirements and ethical standards. The FDL maintains comprehensive records of these decisions and their outcomes, supporting both compliance requirements and continuous improvement efforts.

AI also helps preserve institutional knowledge by capturing valuable feedback from experienced personnel. As seasoned team members move on, their decisions, adjustments, and insights don't disappear. Their applied wisdom is recorded, refined, and reused. This evolving knowledge base becomes a practical guide for new employees, supporting consistent operations and accelerating learning across the organization.

Integrating AI into data management shifts the role of technology from being a supporting tool to an active contributor by building a living system that learns from every result, adapts with every input, and improves through every decision.

Insights in Action

Norman was halfway through his morning coffee when his tablet chirped. He'd been skeptical of the new tool when IT handed it out, but he had to admit, it was starting to grow on him.

"Alert: Anomaly detected in Assembly Line 3 output," the screen flashed.

Norman frowned. He'd been keeping an eye on that line, but everything had seemed fine. He tapped the alert, and a wealth of information sprang to life: operator staffing, production rates, quality metrics, and machine performance data.

"What's this?" he muttered to himself. The overall production rate was steady, but the quality metrics, although within limits, showed a subtle downward trend over the past week. It was so gradual, it would have been easy to miss until it was too late, without the new pattern recognition AI that was rolled out.

Norman went over to the line, tablet in hand. He pulled up the machine settings and compared them to historical data. "Aha!" he exclaimed. The pressure on the main hydraulic press had drifted slightly out of optimal range.

He made a small adjustment and watched as the next batch of parts came through. His tablet quickly displayed the quality metrics improving in real-time.

Satisfied, Norman opened the maintenance app on his tablet and logged the adjustment. As he did, a message popped up: "Similar drift detected on Assembly Lines 5 and 7. Would you like to apply the same adjustment?"

Norman grinned. "You're learning fast, aren't you?" he said to the tablet.

He approved the changes and watched as the system updated the setpoints for the other lines.

As he walked back to his station, Rita passed by. "How's the new system working out, Norman?" she asked.

Norman held up the tablet. "You know, Rita, I think me and this fancy AI are going to get along just fine. It's like having a young apprentice who never sleeps and has a perfect memory."

Rita laughed. "Glad to hear it. Remember, the system learns from you too. Your expertise is making the whole company smarter."

Norman nodded, a twinkle in his eye. "Well then, let's see what other tricks we can teach it. I've got a lifetime of knowledge to share, and it looks like your system is ready to learn."

Chapter 5

The Symphony of Systems

||||||||||||||||||||||||||||||||||||||||||||||||||||||||||||||||||||||||||||||||||||||

Introduction

On the factory floor, an intricate performance is constantly in motion. Machines operate with precise rhythm, conveyor systems move materials in tightly timed sequences, and quality sensors capture thousands of measurements each minute. Each component performs its individual role, yet collectively they form a coordinated system that delivers far more than the sum of its parts.

Like a grand symphony, manufacturing excellence depends on perfect timing and coordination. When one instrument falls out of tune or loses tempo, the entire performance suffers. A slight delay in material flow might ripple through production schedules. A small quality variation could affect downstream processes. Temperature changes in one area might influence product characteristics in another. Each system influences the next, forming a web of interdependencies that must work in concert to achieve goals.

Smart technology acts as the conductor in this performance, keeping time while actively listening, learning, and guiding. It might notice that a precise combination of temperature, humidity, and speed on a piece of equipment consistently produces the best outcomes. Or that scheduling maintenance at just the right moment extends equipment life without impacting production. These insights help fine-tune the operation, aligning every system for optimal performance.

But factories face challenges that symphonies never encounter. Manufacturing runs continuously, with changing materials, conditions, and demands. There's no pause between performances to recalibrate. The orchestra must maintain harmony while adapting in real-time. When disruptions occur, the goal is to adjust quickly and seamlessly, preserving the rhythm without missing a beat.

Modern AI gives manufacturers capabilities far beyond any conductor's baton. These systems monitor every section of the orchestra, detecting subtle signals that might escape human notice. A faint vibration in one machine might predict a downstream quality issue, like hearing a violin slightly off-pitch before it affects the entire string section. Variations in material properties might impact the timing of subsequent steps, requiring the conductor to adjust the tempo across multiple instruments simultaneously.

Understanding how different systems impact one another leads to deeper insights. Instead of waiting for instruments to fall out of tune, smart conductors can spot potential issues early and guide adjustments to prevent them. It's a shift from asking "What went wrong?" to exploring "How can we perfect this performance?" Every decision, whether it's the sequence of orders, timing of maintenance, or adjustments for material changes, has ripple effects that either enhance or disrupt the symphony.

This orchestrated approach unlocks extraordinary possibilities. Production schedules adapt like musical arrangements, flowing seamlessly around equipment maintenance and material availability. Quality control becomes preventive rather than reactive, catching variations before they throw off the performance. Supply chains coordinate like synchronized musicians, each playing their part in perfect time with the others.

Manufacturing excellence comes together one note at a time, with each improvement adding depth and strengthening the harmony between systems. This chapter explores how to become the conductor of your own manufacturing symphony by understanding the connections, reading the rhythm, and guiding every element to perform in perfect coordination.

Mapping the Manufacturing Ecosystem

The first step to understanding the relationships of systems is to document what's already in place by diagramming the manufacturing landscape. Begin with the core production systems, then build out gradually as more connections are revealed. This map becomes a valuable tool for understanding the data environment and spotting opportunities for improvement.

Start by taking a high-level view. At the enterprise level, systems like resource planning and supply chain platforms connect the factory to the larger organization. These systems manage orders, inventory, and deliveries, setting the stage for everything that happens on the factory floor. They shape production priorities and logistics decisions, bridging global strategies with day-to-day operations.

From there, shift focus to the site level, where systems like Manufacturing Execution Systems (MES) and Data Historians play a critical role. MES tracks daily production schedules, monitors progress in real-time, and bridges the gap between enterprise plans and shop floor execution. Historians collect and store time-series data from sensors and control systems, providing insights into performance trends.

Finally, zoom into the factory floor itself. Here, the ecosystem comes alive with activity. Sensors monitor equipment, quality systems verify alignment to standards, and PLCs synchronize their efforts to keep production moving. Each component generates valuable data, but knowledge comes from understanding their interactions. Quality readings might flag equipment wear, or production logs could reveal delays tied to material flow. These connections are where meaningful insights emerge

Mapping is more than just drawing connections between machines and systems. This process includes capturing the story of manufacturing from raw materials to finished goods. Which systems govern each step? How does information flow between them? What causes changes in one part of the system when something shifts elsewhere? Answering these questions reveals how processes work together and where they might fail.

Dig deeper to uncover connections that aren't immediately obvious. Maintenance schedules, for example, might influence product quality in subtle ways, or environmental conditions in storage could affect how materials perform during production. These less visible links may hold the key to meaningful improvements that go unnoticed without a broader view.

A complete map requires input from all sides. Operators understand how machines behave under pressure. Maintenance teams spot patterns in wear and downtime. Quality experts see how small changes affect downstream results. Data moves in both directions, though, so it's just as important to map how factory events influence enterprise systems. A delay on the shop floor can disrupt order schedules managed by planning tools. A rise in defects might point to supplier processes that need adjustment. Capturing these connections helps reveal how changes at the factory level ripple across the entire operation.

Modern tools make mapping these connections easier and more effective. Some companies rely on flowcharts to outline how information moves between systems, while others build digital models that simulate their entire operation. The specific tool matters less than how clearly it shows relationships. A good map should make it easy for anyone to see how their role fits into the larger picture and how their work connects with others.

To build this view, it helps to understand the core systems involved and the roles they play:

- **Enterprise Resource Planning (ERP)**: Manages business-wide functions like order processing, inventory, procurement, and financials, serving as the backbone that connects enterprise operations with production needs.

- **Scheduling Software**: Manages production schedules, allocates resources, and ensures workflows stay on track.

- **Warehouse Management Systems (WMS)**: Track raw materials, finished goods, and inventory levels to ensure a steady flow of inputs and outputs.

- **Quality Management Systems (QMS)**: Track measurements, ensure compliance with product specifications, and flag anomalies during production.

- **Laboratory Information Management Systems (LIMS)**: Manage sample data, ensure lab workflow efficiency, and maintain compliance with regulatory and quality standards.

- **Manufacturing Execution Systems (MES)**: Bridge planning and production by tracking real-time progress, coordinating workflows, and ensuring efficiency across the shop floor.

- **Energy Management Systems (EMS)**: Monitor and optimize energy usage across equipment and processes to reduce costs and environmental impact.

- **Maintenance Platforms / Asset Management**: Monitor equipment health, schedule preventive repairs, and log machine performance data.

- **Data Historian Systems**: Capture and store time-series data from sensors and control systems, enabling analysis of long-term trends in performance, energy usage, and process stability.

- **SCADA Systems (Supervisory Control and Data Acquisition)**: Provide real-time monitoring and control of machinery and processes.

For each system, consider what data it relies on and what it contributes to the ecosystem. Scheduling software needs real-time feedback from production lines to adjust plans as conditions change. Quality systems rely on sensor data to assess product characteristics, while maintenance platforms depend on equipment logs and operator input to predict failures. This step creates a detailed map of how information flows.

Finally, revisit the connections between these layers. How do enterprise systems influence site-level tools? How do decisions on the shop floor ripple back to the rest of the organization? A clearer picture of these interactions helps identify gaps, refine processes, and uncover opportunities for improvement across the board.

Once the basics have been outlined, dig deeper into how these systems interact. What happens when production schedules change? Which systems need updates immediately, and how quickly does that data need to move? Consider what might happen if a system struggles to share information or if delays occur. These questions help identify areas where improving connections could reduce inefficiencies or prevent downstream issues.

Older systems can add complexity to this process. Legacy equipment or software might not integrate easily with modern tools, which can require manual steps to move data. These systems may rely on outdated formats that create bottlenecks in the flow of information. It's important to include these challenges in the map, as they can highlight where targeted improvements or upgrades will have the greatest impact.

Timing is another critical element. Some systems, like SCADA, provide constant updates, while others, like inventory management, might only refresh once per shift. Understanding the timing and frequency of data sharing helps pinpoint where faster communication could make a difference. For example, live scrap counts might allow production teams to adjust processes in real-time, reducing waste or defects.

The map should also spotlight areas where information flow breaks down. Are inventory updates delayed, causing production hiccups? Do quality teams receive equipment data too late to act? These gaps lead to larger issues that negatively impact the rest of the company. Highlighting them on the map helps prioritize fixes and focus efforts where they'll make the most impact.

It's important to keep in mind through this process that manufacturing is constantly evolving. New equipment, process changes, and shifting customer demands mean the map will need regular updates to stay accurate and useful. Don't aim for perfection from the start. Instead, create a flexible map that grows with the company and helps everyone understand how their role fits into the bigger picture.

With a clear map in place, decisions become faster and more informed. If quality issues arise on a production line, the map helps trace connections across systems to identify likely causes. It might reveal shifts in maintenance timing, changes in material properties, or environmental factors influencing performance.

Mapping also brings hidden opportunities to light. It may reveal duplicate data being collected across systems in slightly different ways. Consolidating these efforts can reduce errors and save time. The map might also highlight gaps such as areas where adding sensors could provide valuable insight into process performance. As improvements are planned, the map serves as a practical guide. It can show where an AI tool might drive the most value, where faster responses depend on stronger connections, or how a small change in one area could reduce waste across others.

This tool empowers both the workforce and the automated systems that support them. AI models and other technologies rely on clear context to make meaningful decisions, and the map provides that structure. With a full view of how different parts of the operation connect, these systems can suggest adjustments that enhance overall efficiency rather than just improving isolated processes.

As AI becomes a more integral part of manufacturing, the map grows even more valuable. This technology thrives on well-organized data and defined relationships, allowing it to analyze complex interactions and uncover hidden patterns. With a clear picture of current connections, a trained model can help identify where the latest technologies can deliver the most impact. It helps prioritize upgrades and implement solutions that strengthen the existing setup.

Sharing the map across teams invites a broader range of insights. Operators might spot patterns that others miss, engineers could identify ways to simplify hand-offs between systems, and maintenance staff may suggest smarter coordination strategies. The more people contributing, the better the outcomes.

This visibility strengthens the foundation of a powerful data architecture. When relationships between systems are clearly mapped, AI models can learn faster, make more accurate predictions, and offer insights that truly add improvements to how things run. A well used map becomes more than a reference, it becomes a driver of both human action and intelligent automation.

Modeling Data Relationships for Causal AI

With systems now connected and speaking the same language through the Unified Data Layer, the next step is to understand how that information connects. Gathering data in one place is important, but the real value comes from seeing how each element influences the others.

Causal AI brings a new level of clarity to the data. While traditional models spot patterns, they rarely explain why those patterns exist. Causal AI goes further. It looks for the actual reasons behind what's happening and predicts what could change as a result.

In traditional AI, patterns and associations are discovered, but these findings may lack the context needed for decisive action. Causal AI, by contrast, strives to explain why a pattern occurs, enabling applications to predict outcomes, recommend interventions, and simulate scenarios with a much higher degree of precision. It moves beyond "What's happening?" to "Why is it happening, and what will change if an adjustment is made?"

This approach also helps anticipate outcomes before making changes. Consider adjusting the speed of a production line. It might seem like a quick way to increase output, but understanding the broader system reveals how that speed change could affect downstream quality or create excess inventory. Data models that capture relationships across processes give a clear way to predict and balance these impacts, allowing users to make decisions with confidence.

These insights grow stronger over time. As the facility evolves with new equipment, changing materials, or shifts in demand, the system continues to learn. It refines its understanding of the unique processes, ensuring that recommendations stay relevant and actionable even as conditions change.

Let's break it down with an example. Imagine two issues arise on the factory floor. Machine A starts vibrating slightly more than usual but remains within tolerances. Around the same time, quality defects begin appearing in products processed on Machine B. At first glance, these seem unrelated. Traditional methods might isolate each issue, testing fixes independently or attributing the defects to operator error or material inconsistencies.

An intelligent system that understands relationships sees more than isolated events. It recognizes that two machines share a conveyor and picks up on the subtle vibrations from one causing slight misalignments. These small shifts increase tension, which then disrupts the performance of the second machine. It may need to even go deeper by linking the problem to a specific supplier whose materials tend to be more brittle, showing how the issue only occurs under certain conditions. This changes how the problem is solved. Rather than adjusting Machine B repeatedly, the focus shifts to aligning the conveyor, refining the settings on the first machine, or working with the supplier to improve material consistency. By addressing the real cause, the team avoids unnecessary downtime and improves reliability.

These insights go beyond solving a single issue. Recognizing how material properties affect performance helps guide smarter purchasing decisions. A slightly more expensive material that performs consistently can reduce defects and extend equipment life, saving time and cost in the long run. This broader view leads to improvements that build on each other. Fixing a conveyor might solve a quality issue while also increasing equipment uptime.

As the system gains experience, it learns to predict what will happen next. It recalls that in certain conditions, adjusting conveyor alignment improves results or that pairing specific materials reduces errors. When teams explore new production sequences or tweak machine settings, the system can suggest likely outcomes based on past patterns.

The more connected the data, the greater the impact. Linking quality results with production conditions might reveal inefficiencies in material handling. Logistics patterns might show how small delivery changes improve flow. By connecting insights across systems, operations move from reacting to problems to resolving their underlying causes.

When a product fails to meet quality standards, for example, the cause is rarely simple. It's often the result of several interconnected factors spanning equipment, materials, processes, and people. Pinpointing the root cause requires access to data from multiple sources and the tools to make sense of this interconnected web of data. By connecting systems and surfacing relationships that aren't immediately obvious, AI helps teams see the full picture faster and take action with confidence.

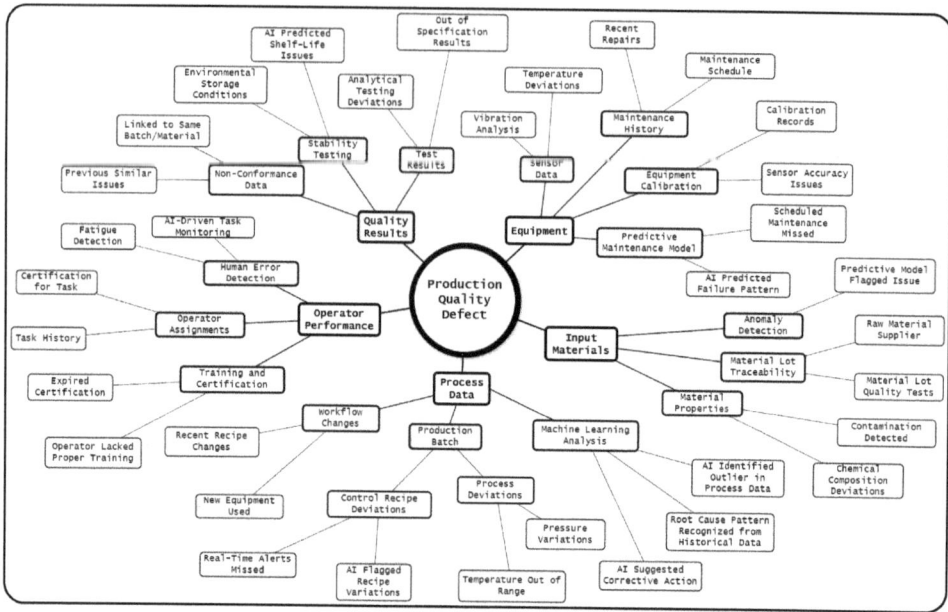

Figure 5-1 Production Quality Defect Root Cause Analysis

The diagram above illustrates how a single production quality defect can arise from a network of interconnected factors across manufacturing. Equipment performance, including sensor readings and maintenance records, plays a critical role. Raw material properties such as chemical composition or physical consistency, add another layer of influence. Process settings like temperature, pressure, or speed directly affect outcomes. Operator actions, including adherence to procedures or recent training, also contribute. Lastly, quality metrics provide key insights into the results of these combined factors. Each branch on the diagram represents one of these potential contributors, highlighting the complex interactions that shape production outcomes.

In this example, an AI solution could pinpoint that a combination of slight temperature fluctuations in equipment, subtle chemical inconsistencies in raw materials, and an operator missing a certification all played a role. This ability to uncover patterns across diverse data streams is what sets advanced analytics apart, turning seemingly unrelated details into actionable insights.

Getting started requires a solid framework with clearly defined connections.

AI-powered tools help by analyzing vast datasets to find hidden relationships and patterns. They might reveal something unexpected, like a specific product line's quality being sensitive to the factory's building temperature. A small discovery like this could lead to significant savings by reducing waste or improving efficiency.

AI doesn't stop at correlations. Techniques like causal inference and deep learning go another step further, uncovering the real cause-and-effect relationships in processes. These insights power predictive maintenance, optimize processes dynamically, and enable systems to make informed decisions automatically.

Once potential relationships across systems are aligned, the next step is to explore how those relationships interact in more complex ways. Graph databases offer a powerful tool for this by modeling how different pieces of information relate to one another across domains. In a graph, each node represents something specific, like a machine, a batch, or a downtime event. The edges between them define how those things are connected. Over time, this network becomes a living representation of how the business behaves.

The strength of this approach comes from the foundation already built. The Unified Data Layer and the Common Data Models organize the raw data into consistent, reusable structures. These models give meaning to the data by defining what each element represents and how it should be interpreted. Without this clarity, relationships can become tangled or misleading. But with these models in place, graph databases can confidently link data from across the enterprise in a way that makes sense.

Graph structures thrive on this consistency. They extend beyond direct links defined in the original models and help uncover second or third level relationships that may not be obvious. For example, maintenance records might not directly reference product quality, but by tracing how equipment performance affects production timing, and how that timing impacts curing or handling, a connection begins to form.

This kind of modeling also supports scalability. As more data flows into the system, the graph doesn't need to be redesigned. It simply grows, adapting to new patterns and inputs. That flexibility is especially valuable when AI is involved. As intelligent systems find new relationships, those discoveries can be folded back into the graph, making it smarter over time.

With this setup, teams can ask more meaningful questions. Which process steps are most sensitive to temperature changes? What sequence of events usually leads to a breakdown? How do delays from a specific supplier impact production and lead to defects? The answers appear through the structure of the graph both visually and analytically. As AI identifies deeper patterns, these relationships evolve, helping systems adapt and processes improve.

Causal AI leverages the paths between the graph database nodes to uncover true deterministic relationships. This allows it to provide actionable insights by explaining why those relationships exist and predicting how adjustments will impact operations. This turns the technology ecosystem into an evolving network where every connection has a chance to tell a story.

Creating Value from Time-Series Data

Within a factory there are multiple streams of data each having their own rhythm. Sensors might record thousands of readings during a single shift, offering granular details about every fraction of a second of the process. In contrast, operators record key events at specific moments such as marking a batch start, logging a material change, or performing quality checks. Broader reports summarize performance across multiple shifts, weeks, or even months, revealing overarching patterns. These layers of data are interconnected, but their differing timescales and purposes make it challenging to align them effectively.

Context turns raw data into meaningful insight. A temperature spike might seem minor on its own, but its relevance becomes clear when linked to a batch that later failed a quality check. A small note in a maintenance log could explain gradual changes in product consistency. Each stream of sensor readings should connect to specific work orders, batches, or equipment states. Tagging data with labels like "mixing" or "heating" makes it searchable and comparable across runs. These markers help teams find what they need faster and understand the conditions behind certain outcomes. With context in place, raw sensor output becomes a valuable tool for smarter decisions.

The challenge comes from bringing together data that moves at different speeds and levels of detail. A single batch might produce tens of thousands of sensor readings, while only a few operator actions are recorded during that same time. Reports may cover dozens of batches, offering a much broader view. To understand how these layers influence each other, they need to be connected in a way that shows their relationship clearly.

A practical way to connect sensor data with process insight is by grouping it into meaningful time intervals. Rather than reviewing each individual reading, data can be aggregated into summaries, such as five minute averages, that reveal broader trends. In some cases, tracking the highest and lowest values during a batch run provides a clearer picture of variation and risk.

Another useful method is to align data with key events. When an operator logs a batch start or a material change, capturing sensor readings just before and after that moment offers a focused view of the surrounding conditions. These snapshots provide context, making it easier to understand what was happening during critical stages of production. With this approach, the data begins to tell a clear story that links actions to outcomes.

Thresholds can also serve as natural markers by tying real-time data to significant moments. For example, monitoring for temperature spikes or pressure drops outside acceptable ranges can highlight where issues might be emerging. These markers flag potential problems while creating clear links between streaming sensor data and the production records that document what was happening at the time. This makes it easier to identify the root cause of an issue and understand how it impacts the broader operation.

To uncover meaning in detailed sensor data, summary statistics help surface useful patterns. Instead of reviewing every reading, teams can focus on metrics like averages, variability, or how conditions shift over time within each batch or production run. These summaries create a link between granular sensor input and higher level production records. They reveal trends, highlight anomalies, and make it easier to spot unusual results that might otherwise be missed.

Streaming data becomes especially powerful when tied to maintenance records. Linking vibration, temperature, or speed data to repair logs can reveal patterns that point to early signs of wear. Recurring signals may indicate upcoming failures, while specific combinations of operating conditions might explain faster degradation.

This same approach supports process improvement. For energy management, time-stamped data from meters offers a live view of how power is used across shifts, machines, and processes. When this data is mapped to production schedules and equipment settings, inefficiencies start to appear. It may reveal that energy-heavy systems are running during low-demand periods or that certain configurations use more power than necessary. These insights lead to adjustments that lower costs and support sustainability goals.

Time-series data also plays a key role in improving quality. When detailed process conditions are compared across many runs, clear patterns emerge. By connecting specific settings or environmental factors to performance outcomes, teams can identify which conditions produce the best results and repeat them with confidence.

Effectively managing streaming data begins with prioritization. Sensors and programmable controllers generate an endless flood of information, but not every data element demands the same level of attention. Detailed data should be preserved for critical events, anomalies, or key process transitions where understanding specific conditions is crucial. During normal operations, summaries and averages are sufficient to capture trends without overwhelming storage or analysis capabilities. This approach ensures the data remains actionable and focused on driving improvements.

Automation plays a key role in maintaining this clarity and efficiency. Operators and production teams need to focus on running the factory, not managing data entry. Automated systems can take on the task of linking sensor readings with batch records and tagging them with appropriate context. This ensures a seamless integration of real-time data with transactional records, giving a unified view of the operation without adding extra workloads for teams.

Consistent handling of time-series data ensures that summaries are calculated in the same way, event triggers respond as expected, and patterns can be compared with confidence. The Common Data Model plays a key role in aligning these values with their meanings. The Unified Data Layer then brings this information together, integrating data from across systems and adding the context needed to understand it.

This combination of structured and time-series data forms the base elements of the Unified Manufacturing Data Architecture. By linking streaming data with batch records or job orders, the architecture builds a connected environment where every input has a place in the bigger picture.

This consistency is essential for AI. With context-rich time-series data, intelligent systems can identify patterns across shifts, equipment, and product lines. Whether optimizing energy use, predicting equipment failure, or improving product quality, AI depends on this clean, well aligned data to deliver accurate insights. This allows time-series data to elevate to another key enabler of operational intelligence.

Weaving Data with Digital Threads and Digital Twins

Two distinct but complementary technologies are now changing how manufacturers manage and understand their operations. Digital threads track the complete history of products through their lifecycle, while digital twins create living models of current operations. Though often confused, these tools serve fundamentally different purposes. One records where things have been. The other shows where they are now.

Digital threads create an unbroken record of a product's journey from design through delivery. Think of them as the manufacturing equivalent of a genealogy tree, tracing every decision, change, and event that shaped the final outcome. This historical record connects design files, production settings, quality measurements, and supplier information into one continuous narrative.

When an engineer modifies a design, that change becomes a part of the thread and automatically updates production requirements, quality specifications, and supply chain orders. Teams can trace any quality issue back through the entire production history, identifying exactly when and where problems originated. A defect discovered during final inspection might reveal its roots in a material substitution made weeks earlier or a design compromise from the initial concept phase. Patterns emerge from the accumulated history of thousands of products. Materials that consistently produce superior results become obvious. Design choices that lead to production challenges reveal themselves. The thread captures not just what happened, but why it mattered.

Consider a manufacturer discovering that products made with a specific alloy consistently outperform others. Without a digital thread, this insight might stay buried in quality reports. With the thread in place, the discovery flows back to design teams, procurement specialists, and production planners. Future products benefit from past experience automatically.

The power of digital threads grows when combined with AI. Machine learning algorithms can analyze thousands of product histories simultaneously, spotting correlations people might miss. They might discover that products manufactured during certain temperature ranges show fewer defects, or that specific supplier batches correlate with higher customer satisfaction. These insights, grounded in actual product genealogy, drive continuous improvement across the enterprise.

While digital threads record history, digital twins model the present. These virtual replicas mirror physical assets, processes, facilities, or even the entire supply chain in real-time. Fed by continuous sensor data, operational events, and environmental conditions, twins evolve moment by moment alongside their physical counterparts.

A digital twin of a production line shows current machine states, active product flows, and real-time performance metrics. It reveals bottlenecks as they form, predicts quality issues before they occur, and suggests optimizations based on current conditions. Unlike static models or historical reports, the twin lives and breathes with the operation it represents.

Teams use digital twins to experiment without risk. If they want to know how changing machine speeds affects quality, they'll run the scenario in the twin first. If they're curious about rearranging the production sequence, they can test it virtually before moving a single piece of equipment. The twin provides a safe environment for exploration, learning, and optimization.

Maintenance teams particularly benefit from this real-time modeling. A twin might show that a critical bearing runs slightly hotter than usual, even though it remains within specifications. By comparing current behavior to the twin's accumulated knowledge of normal operations, teams can spot problems while they're still minor and schedule repairs during planned downtime.

Digital twins provide an ideal environment for AI-driven optimization. By continuously mirroring real operating conditions, they give AI models a realistic space to learn and improve. Algorithms can run thousands of virtual trials, exploring different ways to refine production schedules or enhance quality control. The most effective strategies can then move straight from simulation to the factory floor.

The relationship between AI and digital twins creates a powerful feedback loop. As AI makes predictions and recommendations, real-world results flow back through the twin, improving future accuracy. Each interaction makes both the AI and the twin smarter, creating a system that learns and adapts continuously.

Digital threads and digital twins gain their power from the unified architecture that supports them. The following diagram shows how these components fit within the broader system:

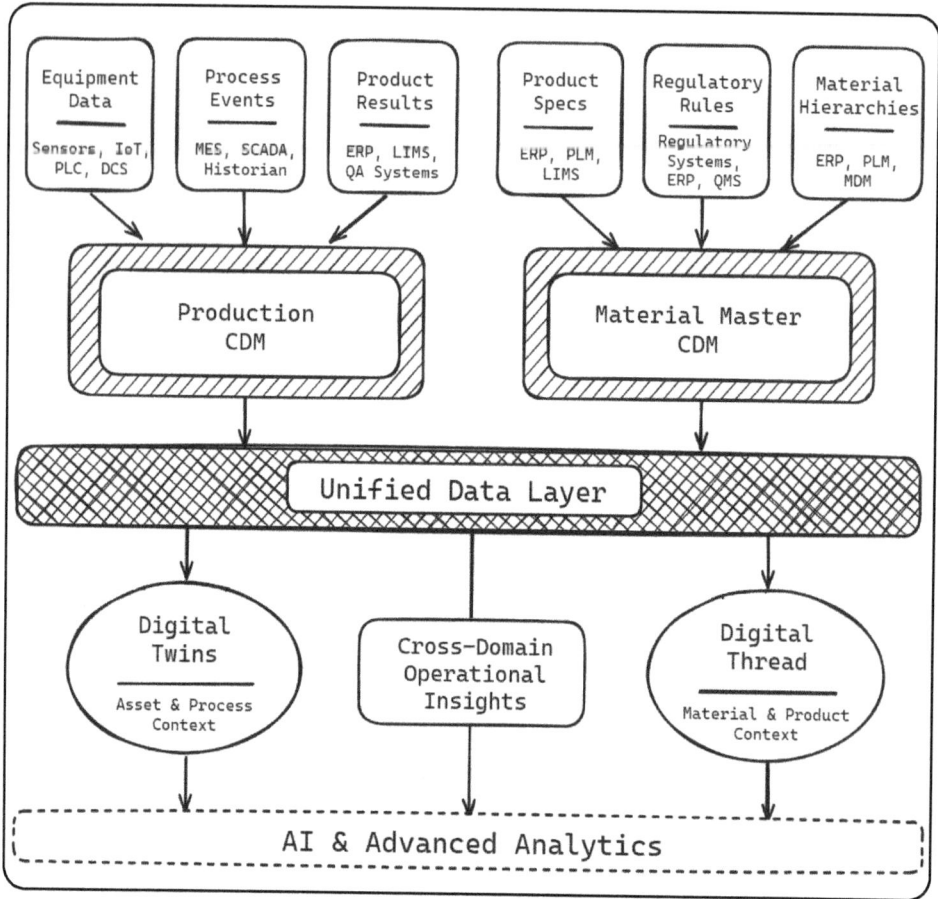

Figure 5-2 Digital Twins and Digital Thread in the UMDA

Operational data flows from the factory floor through the Production Common Data Model, carrying equipment readings, process events, and quality measurements. Reference data moves through the Material Master Common Data Model, providing specifications, standards, and material properties. Both streams meet in the Unified Data Layer, where they combine into a complete operational picture.

From this foundation, digital threads capture the historical journey of materials and products, while digital twins model the current state of assets and processes. The thread provides context by showing how things reached their current state, while the twin provides insight by showing what's happening right now and what might happen next.

This separation of concerns prevents confusion and strengthens both tools. Threads don't try to model real-time operations. Twins don't attempt to maintain complete historical records. Each focuses on what it does best, drawing from the same Unified Data Layer but serving different operational needs.

Together, digital threads and twins create an environment where AI can truly understand manufacturing operations. When a quality issue arises, AI can use the thread to understand the complete history of affected products while simultaneously using the twin to adjust current production. When planning process improvements, AI can study successful patterns from the threads while simulating potential changes in the twin. This combination of historical insight and real-time modeling enables decisions that balance learned experience with current conditions.

The distinction between these tools becomes clear in practice. A thread might reveal that products manufactured on Line 3 using Material Batch 2B show superior durability. The twin shows that Line 3 currently runs at 85% efficiency due to a minor mechanical issue. AI combines both perspectives, perhaps recommending a temporary shift of high-priority products to Line 3 despite its reduced speed, knowing the quality benefits outweigh the efficiency loss.

Manufacturing excellence requires both memory and awareness. Digital threads provide the memory, recording every decision and outcome for future learning. Digital twins provide awareness, reflecting current reality with perfect clarity.

Viewed this way, these tools aren't technologies just added on top of a data lake. They are natural outcomes of a thoughtful data strategy. Their strength lies in combining operational signals with rich reference data, allowing organizations to ask more complex questions and find answers faster. This architecture makes it possible to treat digital twins as evolving, real-time assets and digital threads as living records that capture the full product narrative.

Leveraging AI and Human Insight to Drive Change

Throughout this chapter, we've built an understanding of how to connect and contextualize manufacturing data. From exploring the relationships between data points to managing live streams and embedding digital threads, these principles lay the groundwork for an intelligent factory. But the biggest benefit of this data architecture is its ability to go beyond solving today's problems. It creates a system ready to adapt, evolve, and thrive in fluid markets.

To illustrate how some of these elements come together to create an AI-driven manufacturing environment with humans and equipment in the loop, let's examine the following diagram:

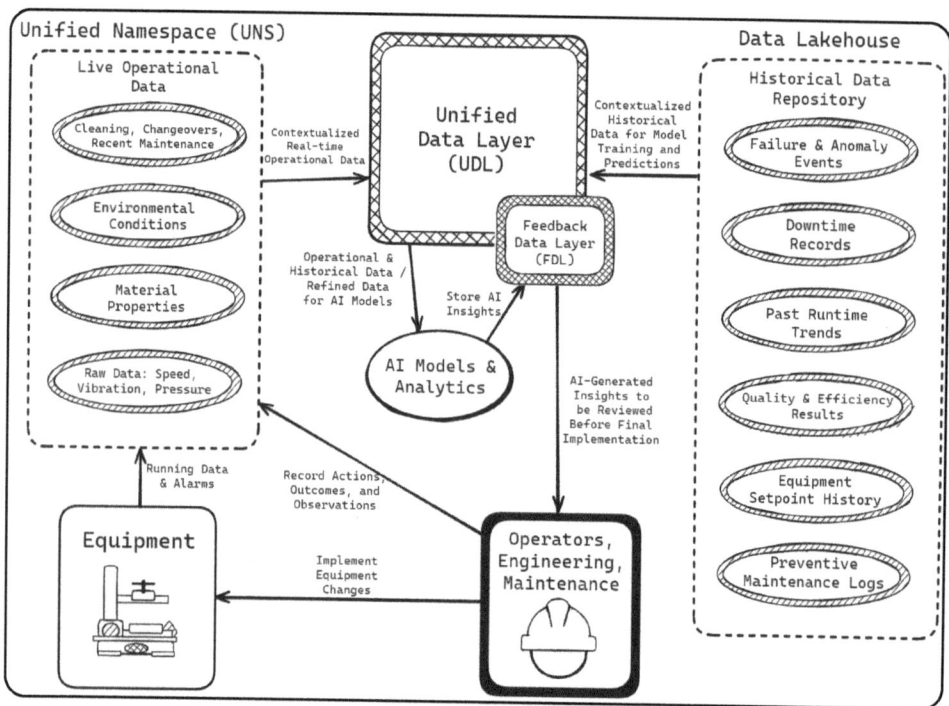

Figure 5-3 AI Insights with Humans in the Loop

This diagram shows components of an example adaptive system and how they interact to drive continuous improvement. At its core, we see the seamless integration of live operational data, historical information, artificial intelligence, and human expertise.

Live data enters the system through the Unified Namespace, pulling signals from sensors, production systems, and environmental monitors. It captures key details like machine status, process conditions, and material characteristics. This stream offers a real-time view of what's happening on the shop floor, helping teams react quickly as conditions shift. The Unified Namespace will be covered in more detail later in the book, but for now, think of it as the central highway for real-time industrial data. While Common Data Models help structure this information, they're left out of this diagram for clarity. Once integrated into the Unified Data Layer, the data gains added context, linking each signal to the right batch, process, or parameter.

Historical data adds depth to this. Maintenance logs, past production records, and quality metrics provide details on the long-term trends and patterns that are essential for informed decision making. Together, live and historical data create a multidimensional view of the operation, enabling teams to address immediate needs while planning for the future.

This data foundation becomes the launchpad for advanced analytics and AI. Algorithms can analyze these rich datasets to uncover hidden opportunities, predict potential problems, and recommend specific actions. AI might identify subtle patterns linking machine vibrations to quality issues, or suggest optimal production sequences to reduce energy use. These insights go beyond what human analysts could discern.

But insights alone aren't enough. The system's strength lies in translating those discoveries into actions that lead to real-world change. The Feedback Data Layer plays a critical role here. It acts as the bridge, capturing AI-generated recommendations and routing them to the people who can evaluate and implement them. Human expertise ensures that changes are considered in the full context of operational realities such as upcoming maintenance, material characteristics, or local environmental conditions that may not be captured.

This collaboration between AI and human teams is what makes the system adaptive. AI excels at processing vast amounts of information and finding correlations. Human experts provide the judgment and experience needed to refine these suggestions and turn them into practical improvements. Together, they form a feedback loop where each implementation adds new knowledge to the system. The AI learns from these results, refining its models to deliver even better insights next time.

This cyclical process drives continuous improvement. Changes suggested by AI are tested and reviewed, with their outcomes informing future recommendations. Operators and engineers gain a deeper understanding of their systems, and AI algorithms grow more accurate and insightful. Over time, this partnership creates a system that anticipates challenges, optimizing processes and uncovering opportunities for improvement.

Adaptive manufacturing does more than improve daily workflows. By building a real-time, comprehensive view of the manufacturing process, organizations become more agile and responsive. They can adjust quickly to market changes, experiment with new production techniques, and refine product designs faster. Flexibility has become essential when customization and speed to market determine success. The ability to reconfigure production lines for new products, fine-tune processes to meet specific quality goals, and scale operations to match fluctuating demand can set manufacturers apart in a competitive landscape.

Looking further ahead, this integrated approach paves the way for autonomous systems that redefine how manufacturing operates. As AI models grow more advanced and teams gain confidence in their capabilities, many routine decisions and optimizations will shift to automated processes. Human expertise will focus on strategic decisions and innovation, leveraging the insights generated by AI-driven systems to shape the future of manufacturing.

This evolution also supports a growing focus on sustainability. Real-time optimization minimizes waste, reduces energy consumption, and ensures resources are used efficiently. Predictive maintenance prevents unnecessary downtime and extends the life of critical equipment. Adaptive production systems can avoid overproduction by responding precisely to demand, reducing surplus inventory and associated waste. Additionally, the data and insights provided by these systems support sustainable product design and lifecycle management, enabling more environmentally responsible manufacturing practices.

The vision of adaptive manufacturing reshapes more than just processes.

It redefines the relationship between people, technology, and data across the entire operation. Change is no longer seen as a disruption, it becomes a driver of growth and learning. Automation takes on repetitive tasks, freeing up time and space for people to focus on what they do best such as solving problems, thinking strategically, and working collaboratively. This partnership between human insight and intelligent systems ensures that experience, creativity, and context remain central to success.

Most organizations start small, with targeted projects that deliver both impact and learning. That might mean synchronizing historical and real-time data on a single line, using predictive analytics to extend equipment life, or tracing product information through digital threads. Each step forward yields value while strengthening the foundation, making future initiatives easier to scale and more effective.

As the system matures, the Feedback Data Layer plays a growing role by capturing results, keeping track of changes, and adding depth to the organization's collective knowledge. This layer turns every improvement into a learning opportunity, helping teams make better decisions with each cycle. What begins as a data project evolves into a self-improving system that becomes more capable over time.

Bringing all these components together creates an operation that does more than react. It anticipates, adapts, and improves continuously. A strong data foundation, paired with flexible architecture and thoughtful implementation, enables teams to take on new challenges, integrate emerging technologies, and maintain high performance.

This chapter has explored how these interconnected systems come together to support more intelligent, responsive manufacturing. Understanding these relationships and building frameworks to support them opens doors to new improvements. When the architecture is well designed and AI is thoughtfully applied, the result is a system that runs with clarity, precision, and purpose. Manufacturing excellence becomes a daily practice, shaped by data and guided by continuous progress.

Insights in Action

Norman was in the middle of his afternoon rounds when his tablet buzzed. He glanced at the screen, expecting another routine alert, but this time it was different.

"Potential supply chain disruption detected. Recommend review of alternative sourcing options," the message read.

Norman furrowed his brow. "Now how in the world did you figure that out?" he muttered to the tablet.

Curious, he tapped on the alert. A web of interconnected data points sprang to life on the screen, showing everything from weather patterns in Southeast Asia to market fluctuations and shipping container availability.

Norman stared in disbelief as he took it all in. The system had correlated a tropical storm brewing near a key supplier's location with historical production data and current inventory levels. It was predicting a potential shortage of critical components in three weeks, long before anyone would have connected those dots.

Just then, Rita rounded the corner. "Ah, Norman! I see you've discovered our new predictive supply chain module. Pretty neat, huh?"

Norman nodded, still scrolling through the data. "I'll say. But Rita, how does it know all this? It's like it's reading tea leaves or something."

Rita laughed. "Not tea leaves, Norman. It's that interwoven architecture we talked about. Remember? They're connecting data from all over; our suppliers, logistics, even global news feeds. The AI then analyzes it all to spot patterns and potential issues."

Norman curiously stroked his beard. "So what do we do with this information?"

Rita pointed to a button on the screen. "See that? It's already generated some recommendations. You can review them and decide which actions to take."

Norman tapped the button, and a list of potential solutions appeared. "Contact alternate supplier in Mexico... Increase buffer stock... Expedite current shipments..." he read aloud. "This is incredible, Rita. It's like having a crystal ball for our supply chain."

Rita beamed. "That's the power of interconnected data, Norman. We're moving beyond reacting to problems and now we're actually starting to anticipate them."

Norman nodded thoughtfully. "You know, when I started in this business, we'd be lucky to get a phone call from a supplier if there was going to be a delay. Now we've got this magic tablet telling us about problems before they even happen."

"It's not magic, Norman," Rita chuckled. "It's just good data management. But I'm glad you're embracing it."

Chapter 6

Balancing Local Control and Global Alignment

|||||||||||||||||||||||||||||||||||||||||||||||||||||||||||||||||||||||||||||||||||||

Introduction

Every manufacturing organization faces the challenge of balancing the unique needs of each facility with the desire for enterprise-wide consistency. No two plants are exactly alike. Differences in products, raw materials, equipment, workforce skills, cultures, and regional regulations shape how each location operates. What works perfectly in one environment might need major adjustments in another. Yet, maintaining consistency in quality, efficiency, and performance across the entire organization remains critical.

Picture this scenario playing out across a global manufacturing organization. A pharmaceutical plant in Switzerland operates under strict regulatory requirements that demand extensive documentation and validation for every process change. Meanwhile, a sister facility in Brazil focuses on high-volume production with different equipment configurations and local supplier relationships. Both plants manufacture similar products, but their operational realities couldn't be more different. Traditional approaches would force both facilities into identical processes, creating inefficiencies and alignment issues.

Now imagine a different approach. The Swiss facility leverages its regulatory expertise to develop enhanced quality monitoring procedures that exceed compliance requirements while improving product consistency. These insights flow through the global network, where the Brazilian plant adapts the quality improvements to fit its high-volume environment. Both facilities benefit from shared knowledge while optimizing for their unique conditions. This isn't wishful thinking. Leading manufacturers are already implementing these adaptive approaches.

Up until now, much of the focus of the book has been on building a foundation for data and AI-enablement by creating standards, defining common models, and developing approaches to unify information across sites. Those elements are essential, but they only tell part of the story. This chapter marks an important shift by recognizing that real-world scenarios require room for local adaptation of global standards. When a facility faces a sudden material shortage, the local team must have the flexibility to respond immediately. When subtle quality issues arise, operators need the ability to adjust processes based on conditions they see firsthand.

The challenge lies in achieving this balance without sacrificing the benefits of standardization. Organizations need frameworks that provide consistency where it matters most while enabling local teams to innovate and adapt. This requires moving beyond the choice between rigid centralization and chaotic decentralization.

Rather than imposing strict, one-size-fits-all rules, organizations can design frameworks that guide facilities toward shared standards while allowing the flexibility needed to adapt. Decision making shifts from being centrally dictated to empowering local teams with the tools, trust, and data they need to solve problems quickly and effectively. Those closest to the work have the clearest view of what improvements are needed. When facilities are trusted to act, they uncover opportunities and solve challenges faster.

This doesn't mean trading structure for mayhem, or autonomy for inconsistency. The purpose is to design systems that enable uniformity where it matters most while leaving space for creativity, responsiveness, and continuous improvement at the site level.

This chapter provides the outline for achieving that balance. You'll discover how to leverage Common Data Models to adapt to local conditions while maintaining enterprise coherence. We'll explore the tools and technologies that enable local data mapping without losing global visibility. You'll learn to optimize AI models for site-specific decision making while preserving the ability to scale insights across facilities. Most importantly, you'll see how the Edge Intelligence Hub creates local autonomy within a globally connected architecture.

Through practical examples and implementation guidance, you'll understand how manufacturers are solving real-world challenges. From specialty chemical producers adapting to altitude differences to food processing plants optimizing for regional preferences, the principles in this chapter enable organizations to harness local expertise while building enterprise-wide intelligence.

The benefits of this balanced approach extend far beyond immediate productivity. Facilities become hubs of innovation and learning. A breakthrough at one site can inspire similar improvements elsewhere, creating a ripple effect of positive change.

As manufacturing grows more interconnected, this balance between local and global needs is no longer optional, it's essential.

The Case for Local Control

Manufacturing takes many forms, reflecting the wide range of products and processes it supports. Each facility operates within a unique environment, shaped by its specific workflows, objectives, and constraints. While this diversity can introduce complexity, it also holds significant potential. When managed thoughtfully, the distinct characteristics of individual sites become a source of strength rather than a challenge.

In a global organization, each site plays a different role. One location may focus on high-volume production, while another handles specialized, low-volume orders. Local regulations, cultural norms, and environmental conditions influence how work gets done, from operating procedures to how data is captured and applied. Each facility brings its own pace, problem solving methods, and data practices to the broader organization.

Forcing a rigid, all-purpose data strategy onto such a diverse network creates more problems than solutions. A system designed for high volume production might slow down a facility focused on customization. Standardization that ignores local nuances can frustrate teams, stifle creativity, and overlook opportunities to make meaningful improvements.

That's where local control comes in. This requires a balance of providing the flexibility to tailor processes to specific needs while maintaining alignment with organizational goals. Local control doesn't mean abandoning consistency, it means recognizing that different sites require different approaches to succeed. Flexibility at the ground level enables better performance across the board.

This adaptability also enhances data quality. When each facility can focus on the metrics most relevant to its operations, the data collected gains greater meaning and usefulness. Teams are more likely to trust information that reflects their actual working conditions.

As an example, consider a manufacturing network with several facilities producing specialty chemicals using similar continuous process equipment. Standard operating procedures are based on conditions at the original site where the process was first developed, including specific temperature and pressure settings. These specifications work well in that environment, but not every location operates under the same conditions.

At one facility situated in a high altitude region, operators find that following the original parameters leads to slower reaction times and lower yields. The local team, familiar with how altitude affects pressure and reaction dynamics, develops a modified approach. They adjust temperature profiles, refine catalyst ratios, and recalibrate pressure settings to better suit the environmental conditions.

In contrast, a coastal site faces a different set of challenges. High humidity leads to inconsistent moisture levels in raw materials and increased wear on specific equipment components. To address this, the local team might implement drying steps, adapt maintenance routines, or fine-tune process parameters to manage the effects of ambient moisture.

Each facility reaches optimal performance by tailoring operations to its environment. Though their methods differ, they both follow a shared framework for collecting and analyzing data by capturing the same key metrics for quality, efficiency, and performance. This consistency allows the organization to compare results across sites, share insights, and build a deeper understanding of what works where.

With this structure in place, the central manufacturing team can support localized improvements without sacrificing standardization. They might even establish a formal methodology for adapting processes to site-specific conditions, helping every facility maintain product consistency while optimizing for local performance.

Operational efficiency improves when local teams can tailor their data practices to fit their workflows, equipment, and specific challenges. Decisions are faster because they're made by the people who know the operation best. Adjustments happen in real-time, keeping pace with changing conditions. Challenges are solved with practical solutions that make sense for that facility, rather than relying on a top-down directive that may miss the mark.

This coordinated approach forms the basis of a network that learns and grows together. Local successes feed into a broader strategy, where innovation and improvement spread naturally. A site that pioneers a better method for energy management can test, refine, and demonstrate its value locally. Once proven, the process can be shared with other locations, adjusted to suit their specific needs, and scaled across the entire organization.

Empowered teams become creative problem solvers, experimenting with new ways to use data to improve processes. One site might develop a method to reduce downtime during changeovers by rethinking how production data is analyzed. Another might optimize energy consumption by correlating equipment performance with environmental factors.

At its core, this approach acknowledges the expertise of the people closest to the shop floor. Operators, engineers, and managers live the processes every day. They know the equipment's quirks, the best workflows, and the nuances of the challenges they face. By trusting them to shape their data strategies within a structured framework, organizations tap into a reservoir of insight that might otherwise go unnoticed.

The aim is to ensure every site can adapt to its local requirements while staying aligned with broader organizational goals. A unified structure supports this alignment, providing consistency across the enterprise as individual facilities tailor solutions to their specific needs. The Common Data Model creates the shared language that makes these local adaptations work in harmony with enterprise objectives.

This flexibility must have defined limits within a structured system. Without clear boundaries, sites risk drifting too far from organizational standards, which can lead to inefficiencies and misalignment. A shared framework, like a production data model, provides this structure. It allows for customization at the local level while ensuring that all sites remain connected to the broader strategy. Standards, clear guidelines, and open collaboration channels keep this balance intact.

Technology is a critical partner in maintaining this equilibrium. AI tools help local teams analyze data, pinpoint inefficiencies, and identify opportunities for improvement. These systems act as bridges between localized practices and global objectives, ensuring that local optimizations contribute to the organization's overall success. They also make monitoring progress easier, providing transparency and consistency across the network.

Local autonomy is a necessary approach that reshapes how organizations achieve growth and resilience. By giving teams the ability to lead change at the site level, companies build a culture of innovation that spreads across global operations.

Striking the right balance between flexibility and consistency turns operational diversity into a strength, enabling each site to succeed on its own terms while still contributing to larger goals.

Challenges of Decentralized Data Management

Giving manufacturing sites the freedom to manage their own data brings both benefits and challenges. When local teams have the freedom to handle information in their own way, they find creative solutions to their specific problems. They work faster and feel more connected to improving their operations. But this flexibility comes with challenges that can ripple across an organization if not carefully managed.

Consider how different facilities might measure something as straightforward as production efficiency. One might focus on output per hour, another on the ratio of good parts to total production, while a third factors in energy consumption or material waste. Each method reflects the unique priorities and processes of that site. Yet, when it's time to compare results or share best practices, these variations make it difficult to see patterns or evaluate overall performance.

The complexity increases when examining how individual facilities collect and manage data. Some sites may use advanced digital tools, while others continue to rely on legacy systems tailored to their equipment and workflows. Local teams shape their data practices around the realities of their operations, factoring in machine capabilities, staff expertise, and daily production needs. While these adjustments can enhance local performance, they may also introduce issues. Inconsistent practices across sites can hinder collaboration and make it harder to adhere to enterprise-wide standards.

A lack of alignment can isolate valuable insights. For example, a facility may develop a new method to reduce waste or improve quality. But if the data behind that success isn't structured in a way others can interpret, the insight remains confined to that location. This disconnect makes it difficult to replicate improvements to other sites, limiting the ability to scale progress and learn from shared experience.

The key is to create consistency where it matters most. Focus on metrics and practices that drive shared goals like quality, safety, and compliance, while allowing flexibility for local teams to optimize their unique operations.

For instance, key performance indicators that influence regulatory compliance or product quality should be calculated consistently across all facilities. This ensures that trends, risks, and opportunities are visible on a larger scale. Beyond these essentials, local teams benefit from the freedom to decide how best to collect and act on the data most relevant to their site's goals. Achieving this balance requires systems that bridge differences without forcing standardization where it's not needed.

Modern data management tools play a vital role in connecting the diverse needs of local facilities with the broader goals of the organization. Advanced systems can bridge the gap between differing data formats and measurement methods, translating local practices into a shared language that supports collaboration and alignment. These tools act as interpreters, allowing each facility to maintain its unique approach while contributing to an integrated organizational strategy.

One of the most significant challenges in decentralized data systems is the risk of creating new silos. When each site manages data independently, information can become fragmented, isolating valuable insights and limiting their reach. Without consistent context, even shared data can be misunderstood or misused, reducing its overall value. Addressing this requires strong tagging practices and clear metadata standards that preserve meaning and support smooth collaboration.

The effects of fragmented data don't stop at internal operations. They can extend outward, disrupting supply chains and weakening customer responsiveness. Without a unified view across facilities, tasks like forecasting demand, managing inventory, and identifying inefficiencies become much harder to execute with confidence.

Adding to the complexity are the demands of data compliance and security. Many industries face strict rules around how data is stored, accessed, and reported. When each facility follows different practices, it becomes more difficult to maintain compliance and more challenging to guard against cyber threats. In this environment, every site becomes a point of vulnerability. Consistent security protocols and centralized oversight are essential to protect systems and meet regulatory expectations across the organization.

Transitioning to a balanced data system goes beyond adopting new tools, it requires rethinking how teams work together and share responsibility. Centralized teams may need time to adjust to a less hands on role, while local teams might initially feel uncertain about handling greater autonomy. Clear communication, tailored training, and a shared understanding of the benefits are crucial to making this shift successful.

Another key consideration is the variation in resources and expertise across sites. Smaller facilities or those with fewer resources might struggle to meet organizational standards or fully leverage their data for improvement. Providing these sites with focused support through training programs, technology upgrades, or access to shared resources will ensure they can contribute effectively to the organization's goals.

The goal is to approach local autonomy with a clear plan for managing its complexities. By anticipating challenges, organizations can design systems that preserve flexibility while ensuring the consistency needed for confident, scalable insights. Giving sites the ability to tailor data practices to their unique environments boosts responsiveness and supports local innovation.

Standards define how data should be structured and interpreted, while tools like APIs and integration platforms make it possible to connect different systems without forcing uniformity. These technologies create a unified operational view while respecting the diversity of local conditions.

Implementing decentralized data management is as much a cultural transformation as it is a technical one. Success depends on building a shared understanding throughout the organization of why both local flexibility and global structure are essential. Leadership must champion the change, teams must trust the process, and systems must consistently reinforce these principles.

When local practices are guided by standard frameworks, the data they produce becomes easier to interpret, compare, and apply across use cases. The next section will go into further detail about how this can be achieved.

Establishing Flexible Data Structures

A properly designed Common Data Model is essential for creating a balanced data architecture that supports both local agility and global consistency. It gives the organization a shared structure for defining core data elements, helping teams across sites speak the same language while allowing room for local adaptation. The model must stay stable at its center but flex at the edges to accommodate real-world differences across facilities.

At its foundation, the Common Data Model establishes a consistent vocabulary and framework. When terms like "component," "assembly," or "batch" are used, their definitions are clear and agreed upon. This clarity reduces confusion, promotes collaboration, and enables data to move more easily between teams and sites. Standardized models simplify enterprise reporting, benchmarking, and cross-site analysis while still giving local teams the freedom to shape the model around their specific needs.

Consider a global manufacturer with facilities serving industries as varied as food processing, pharmaceuticals, and specialty chemicals. At the corporate level, the Common Data Model might define a standard equipment profile for an industrial mixer, including attributes such as Equipment ID, Original Manufacturer, Capacity, and Maintenance Schedule. These shared fields form the baseline across all facilities, ensuring equipment data remains consistent and comparable.

From there, each site can build on this foundation in ways that reflect its unique environment. A food processing plant might add IoT sensor fields to capture temperature, humidity, vibration, and cleaning cycles while focusing on hygiene and maintenance. A pharmaceutical facility could extend the model with attributes related to batch traceability, surface finish standards, and sterilization procedures to support regulatory compliance. A chemical plant might prioritize safety by including pressure readings, corrosion data, and emission monitoring, along with records of safety interlocks and maintenance history.

Each facility works with the same core structure, but extends it based on what matters most in their context. This approach supports local initiatives while providing meaningful enterprise-wide insight. AI and analytics benefit from the consistent structure, making it easier to compare performance, spot trends, and scale solutions. The mixer example, while simple, illustrates how a shared starting point paired with site-specific context allows the organization to grow smarter together by leveraging both centralized elements and local expertise.

Industry standards for different types of manufacturing provide valuable frameworks for guidance. These standards establish consistent ways to define core manufacturing elements such as equipment, processes, and product definitions, while leaving room for site specific customization. For instance, a pharmaceutical plant might adopt the ISA-88 batch production model and enhance it with data fields specific to active pharmaceutical ingredients or sterilization protocols. This tailored approach allows the facility to meet stringent local regulatory requirements while staying aligned with the broader ecosystem.

Achieving this balance takes careful planning and thoughtful governance. Organizations need to determine which elements must be standardized universally and which can be adapted locally. Clear oversight is essential to ensure that site-specific extensions enhance rather than disrupt data integration. Without proper management, customization can lead to compatibility challenges.

A key to achieving this balance is the integration layer that links the standardized foundation of the Common Data Model with locally tailored extensions. This enables information to flow smoothly into an enterprise-wide view, allowing local adaptations to contribute to, rather than disrupt, the organization's broader data architecture.

The benefits don't stop at individual sites. A flexible CDM encourages collaboration by making local innovations available across the organization. For instance, if a food processing facility successfully deploys IoT-based monitoring for mixers, other locations can adopt this data model enhancement, speeding up improvements and promoting consistency. This culture of knowledge sharing turns local advancements into enterprise-wide assets.

Flexible alignment keeps the model useful as technologies shift, business goals evolve, and new opportunities emerge. Extensions to the core structure can be added as needed without affecting the foundation, allowing the model to grow alongside the factory's needs.

To achieve this balance, the architecture must be carefully designed with clear rules for naming, data types, and how elements relate to one another.

These guidelines act as guardrails, helping local extensions fit smoothly into the global model. With this structure in place, data from all sites can fit into the global framework with their local context attached.

Equally important is the governance process. A well managed governance system keeps the model aligned with operational realities while supporting controlled updates and refinements. Regular collaboration between the central data team and local sites ensures that the framework evolves in a way that reflects real-world needs.

Tools and Technologies Supporting Local Data Mapping

The right tools for data mapping play a critical role in empowering factories to operate with local autonomy while staying aligned with standard practices. These tools bridge the gap between site-specific data practices and unified Common Data Models. By focusing on flexible, scalable, and interoperable technologies, organizations can create a data ecosystem that supports diverse operational needs without compromising consistency.

The tools must be flexible enough to handle differences in how sites collect and label their data. For example, one facility might measure efficiency in pieces per hour, while another uses a ratio of good parts to total output.

Scalability is just as important when selecting the right tools. As operations grow, data mapping solutions need to keep pace with increasing volume and complexity. A system that performs well for a few sites should also support dozens, or even hundreds, of locations. The tool must adjust naturally to new data sources, more users, and changing business demands. To achieve this, the mapping process must remain efficient and reliable, no matter how much data demand evolves.

Interoperability is another cornerstone of effective data mapping technologies. Manufacturing data comes from a wide range of systems, from legacy databases to modern IoT platforms. Mapping tools must integrate seamlessly with these diverse sources by extracting and aligning data to the CDM. This interoperability ensures smooth data flow across the organization, creating a unified view of operations without requiring sites to overhaul their existing systems.

In addition, the tools should support full traceability. Every mapped data point needs a clear lineage that shows where it came from and how it was altered along the way. Transparency helps build confidence in the data and allows teams to quickly trace issues back to their source, making it easier to correct problems and improve overall data quality.

Another essential component in this ecosystem is data virtualization, which is particularly important in environments where real-time access is critical, or where regulations restrict the movement of information. By creating a virtual layer that connects local data systems to the CDM, virtualization tools preserve local autonomy while ensuring the organization gains a complete picture.

Adding to this foundation are metadata management systems, which bring clarity and consistency to local data mapping. These systems act as custodians of context, tracking details such as data lineage, definitions, and relationships. In a global manufacturing network where terms and processes may vary between sites, robust metadata management ensures a common understanding of information. Automated metadata capture and mapping reduce the risk of human error and accelerate the onboarding of new data sources. This clarity ensures decisions are based on data that is accurate and also understood by all teams.

Master Data Management (MDM) solutions play a vital role in strengthening the local data ecosystem by ensuring consistency in foundational elements such as product definitions, supplier records, and customer details. These core reference points act as anchors, giving structure and meaning to the data generated across different sites. This adds the context needed to make local data useful at the enterprise level. For example, when two facilities refer to the same part or material in slightly different ways, MDM aligns those references through a common set of identifiers and attributes. This ensures that insights drawn from one site can be accurately interpreted and compared alongside data from another, without confusion or duplication.

To see how these elements come together in practice, consider a manufacturing organization with facilities across different regions. Each site may have developed its own approach to tracking maintenance, resulting in a variety of systems and methods for capturing similar types of information.

One site might use advanced sensors to monitor equipment automatically. Another might depend on time-based scheduling through traditional maintenance software. A third could rely on manual inspections with basic measurements recorded by hand. These varied approaches reflect local needs and available resources, but they also create issues when trying to build a unified view across the organization.

Without a structured mapping strategy, these differences remain locked within each facility. Insights stay local, and opportunities to share best practices or uncover enterprise-wide trends are missed. Implementing mapping technologies can help bridge these gaps through a set of coordinated tools.

For example, integration platforms can pull data from different local systems without disrupting how those systems are used day to day. Metadata tools can connect local naming conventions to a standardized structure, creating a common language between sites. Master data management can define consistent references for equipment types and specifications, while data virtualization tools can present a single view of maintenance information, even when it comes from different formats and systems.

This kind of approach allows each site to keep using the tools and terminology that work best for them, while still contributing to a shared understanding across the organization. Engineers and analysts at the enterprise level gain access to insights that span facilities, making it easier to spot recurring issues or evaluate the effectiveness of different maintenance strategies.

As sites upgrade their systems or bring new facilities online, the mapping framework adjusts alongside them. It absorbs these changes without losing the consistency needed for enterprise-level analysis. The result is a flexible, scalable system that respects local expertise while supporting a broader organizational perspective.

When putting data mapping technologies into practice, the most effective results come from focusing on business value rather than just technical features. The process should begin with a clear assessment of the current landscape. This means identifying where local differences add value and where they create unnecessary complexity. That evaluation helps clarify which local practices should remain and which could benefit from being brought into a more consistent structure.

Active involvement from stakeholders is essential throughout the process. Local teams should be part of the effort to map their data to the Common Data Model, making sure the structure reflects how work actually gets done, not just how it looks on paper. When people on the ground help shape how their data connects to the broader system, they develop a clearer understanding of its purpose and are more likely to support and sustain the changes.

Throughout the process, it is important to avoid the common mistake of forcing too much standardization, which can disengage teams and reduce local effectiveness. The best results come from finding the right balance, standardizing where it supports scale and consistency while preserving flexibility when it strengthens performance.

As we look ahead, the role of Artificial Intelligence will redefine how local data mapping is approached. AI technologies will simplify complex, time-consuming tasks by identifying patterns in local data, suggesting mappings to the Common Data Model, and learning from past actions to enhance future accuracy. Predictive analytics and automation capabilities embedded in AI-driven tools reduce the manual effort required for data mapping, freeing teams to focus on more strategic activities like identifying opportunities and making informed decisions.

For these tools to function effectively within the enterprise architecture, they must integrate effortlessly with the Unified Data Layer by channeling locally mapped data into the centralized framework. Tools equipped with robust APIs and connectors designed to interface with the UDL enable real-time synchronization and smooth data flow, supporting a cohesive data architecture.

By bringing together scalable integration platforms, virtualization tools, metadata management systems, and AI-driven technologies, organizations can build a data ecosystem that is ready for the future. As these tools continue to advance, the ability to map and connect local data will grow with them, making it easier to support new technologies and unlock opportunities that scale over time.

Optimizing AI Models for Local Decision Making

As manufacturing operations continue to become more connected and standardized data approaches are applied, the next step is putting intelligence to work closer to where decisions are made. Facilities need tools that gather and organize information while turning it into immediate, actionable insights. AI models built to run locally provide this ability.

Edge AI is designed to run directly within the factory environment, enabling fast, informed decisions without waiting for external systems. These models are built to operate on standardized data products, ensuring they can be deployed across multiple facilities without needing complete redesigns for each location. While each site may use the models to respond to its own real-time conditions, the consistency in data structure ensures that the same AI tools can be applied broadly, making it easier to scale solutions across the organization.

Making this possible requires AI models that are lightweight, efficient, and capable of running in real-time without depending on constant cloud connectivity. Manufacturing sites need solutions that fit within local computing environments while delivering immediate results. The models must be built on a common framework, allowing them to fit naturally into the Unified Manufacturing Data Architecture and making it easier to apply successful approaches across multiple locations.

Whether deployed in a high-volume assembly plant or a specialized facility producing custom components, AI models must be adaptable and optimized for real-world variability. A well-designed Edge AI solution enables sites to respond dynamically to changing conditions, continuously refining both local and global processes.

Take quality control as an example. An AI model inspecting parts for defects must analyze thousands of images per hour without disrupting production. A poorly optimized model might require too much computing power, slowing down operations or demanding costly hardware upgrades. A well-structured model, on the other hand, extracts only the most relevant features, allowing it to make rapid, accurate decisions while staying within the site's available resources.

Predictive maintenance is another case where speed matters. AI models monitoring vibration or temperature trends must identify patterns fast enough to prevent breakdowns. If analysis takes too long, the warning might come too late, leading to unplanned downtime or lost production.

Designing AI for local decision making starts with building intelligence that is both focused and practical. Models must prioritize the most relevant data, avoid unnecessary complexity, and integrate naturally into site operations. Achieving this requires close collaboration between AI engineers, manufacturing experts, and IT teams to ensure that solutions are effective in real-world environments.

The design for these solutions must include how to make them flexible enough to perform well across different sites without starting from scratch. A model predicting defects at one facility might need slight adjustments when used at another, based on differences in materials, processes, or environmental factors. Instead of creating entirely new models for each location, organizations can build adjustable frameworks around standardized core models that are fine-tuned with local data.

Task-specific design is a critical factor in ensuring AI models remain effective at the site level. A one-size-fits-all model is rarely the best solution, just as using the same process for every production task would lead to inefficiencies. Instead, AI should be tailored for the specific job it performs.

For example, in a chemical processing plant, different AI models might be used for different aspects of operations. Anomaly detection in a reactor might focus on patterns in temperature, pressure, and chemical composition, while a predictive maintenance model for pumps and valves might analyze vibration and flow rates. By designing each model for its specific function, we ensure that AI remains efficient, focused, and capable of running within local constraints.

A practical approach to model design is to start simple and iterate. Instead of attempting to build the most sophisticated AI model upfront, teams can begin with a simplified version that captures the most critical insights and predictions. This iterative strategy ensures AI delivers immediate value while allowing continuous refinement based on real-world feedback. Much like continuous improvement in manufacturing processes, AI models can evolve over time, adapting to changing needs without causing disruption.

Another essential design consideration is ensuring that models can handle real-world variability. Manufacturing environments can be influenced by fluctuations in environmental conditions, material inconsistencies, and operational shifts. AI models need to remain resilient, adapting to these changes without losing effectiveness.

Consider an AI model used for optimizing robotic welding in an automotive assembly plant. Factors like slight variations in metal thickness, changes in ambient temperature, or differences in gas flow rates can all impact weld quality. If the model is too rigid, it might fail to adjust for these real-world fluctuations, leading to defects or inefficient welds. By incorporating adaptive learning mechanisms, the model can continuously refine its predictions based on real-time data, ensuring consistent quality while responding to subtle process changes.

Building flexibility into AI models is essential for balancing local responsiveness with performance. Instead of depending on large, complex models that need significant processing power, teams can use techniques that make AI both smarter and more efficient. Model pruning trims away parts of a model that add little value, helping it run faster while keeping accuracy intact. Quantization simplifies how calculations are performed so AI can operate smoothly even on lower-powered devices. Knowledge distillation takes the insights of large models and compresses them into smaller, focused versions that deliver reliable results with far less computing demand. These methods work together to create AI tools that fit well at the local level, enabling fast, effective decisions right where they're needed.

Consider an AI system used for equipment failure predictions in an automotive parts facility. The original model may analyze vast amounts of sensor data such as vibration, temperature, pressure, and power usage, and use this to detect signs of wear in robotic assembly arms. While this model is highly accurate, it may contain redundant pathways that slow down processing. Through pruning, the model can focus on the most critical data points, such as vibration patterns and temperature fluctuations, while discarding less relevant computations. This results in a faster, more efficient AI system that can identify maintenance needs in real-time without overloading on-site computing resources.

Quantization follows a similar principle by refining the way AI models handle numerical computations. In manufacturing, process optimization involves reducing waste, whether it's excess material, unnecessary motion, or redundant steps in production. The same applies to AI. By lowering the precision of calculations (e.g., from 32-bit to 8-bit), quantization enables models to run faster and more efficiently on embedded devices while still delivering reliable insights.

Take an AI-driven inventory management system that tracks material flow in a metal fabrication plant. The original model may use high-precision calculations to predict raw material consumption, requiring substantial computing power. By applying quantization, the model can achieve nearly the same accuracy while using lower-bit computations, allowing real-time monitoring of material usage without overburdening local infrastructure.

Knowledge distillation builds on these concepts by transferring expertise from a large, complex AI model into smaller, specialized versions. Just as an experienced operator trains apprentices by focusing on the most critical aspects of their work, this technique ensures that leaner AI models retain key decision making capabilities while eliminating unnecessary complexity.

Imagine a pharmaceutical plant using a high-level AI system to oversee the entire production process, from ingredient mixing to packaging. This model tracks thousands of variables to maintain product consistency. Running such a large model on every machine would be impractical. Through knowledge distillation, task-specific models can be developed. This may include one focusing on ensuring proper ingredient mixing, one monitoring sterilization processes, and another optimizing packaging line efficiency. These smaller, trained models enable real-time decision making at each stage of production while staying aligned with quality and compliance standards.

When manufacturers design AI models with a modular, task-specific approach built around standardized data, they create intelligence that scales effectively across operations. These refined models deliver insights precisely where and when they're needed, eliminating unnecessary complexity without sacrificing accuracy. The iterative development of these models allows them to continuously improve while adapting to the natural variability of production environments. This balanced approach advances AI from a centralized resource into a practical local tool built to enable site-level needs.

Achieving Local Autonomy with the Edge Intelligence Hub

The Edge Intelligence Hub (EIH) takes the promise of Edge AI and transforms it into a true system of action built to enable factories. This means giving each site a clear framework for how local intelligence and enterprise goals stay in sync. The EIH sets the rules of engagement such as what decisions operators can make instantly, what insights need to roll up for broader visibility, and how local and enterprise systems collaborate. By building structure into the edge, manufacturers move beyond scattered Proof of Concept tools and create a dynamic, coordinated network that empowers sites to act quickly.

The EIH processes and analyzes data in real-time, enabling rapid, context-aware decision making. Instead of sending every data point to a central system, the EIH applies localized analytics to identify issues, optimize processes, and implement changes immediately. Whether adjusting equipment parameters, scheduling maintenance, or fine-tuning quality control, it empowers sites to stay agile and efficient.

The EIH acts as a local data management center, fully integrated within the broader manufacturing network. It continuously processes information from sensors, quality control systems, operational metrics, and other local systems, generating actionable insights tailored to each site's unique conditions. For example, if an equipment sensor detects a sudden spike in vibration, the EIH can assess the situation within seconds. It considers factors like the equipment's maintenance history, the demands of the current production batch, and environmental conditions. Based on this real-time analysis, it can recommend immediate adjustments to prevent quality issues or equipment failures.

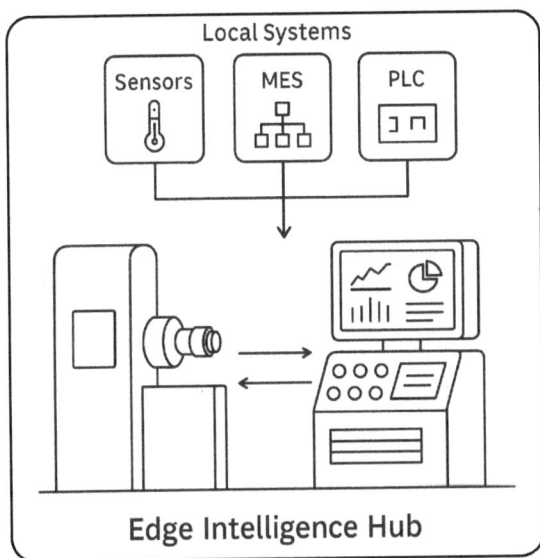

Figure 6-1 The EIH Reacts to Live Data from Local Systems

It's important to note, the EIH doesn't operate in isolation. It bridges local autonomy with enterprise alignment, ensuring decisions remain consistent with global standards while giving each site the flexibility to adapt to its specific challenges. Standardized models, corporate best practices, and shared goals guide the EIH. As outcomes are realized, local insights feed back into the UDL, strengthening the organization's overall knowledge base.

Factory operations teams benefit from broader enterprise expertise while contributing innovations shaped by real-world conditions. For instance, if a facility develops an energy optimization strategy tailored to its unique equipment setup, the EIH can validate and structure those findings for broader application across the network.

By embedding real-time analytics and decision making capabilities directly at the site level, each location becomes a hub of insight and innovation, responding dynamically to operational demands while contributing to the larger organizational strategy.

Achieving localized insight requires systems that can understand, interpret, and act on information within the flow of operations.

At the core of the EIH is the intelligence layer. Here, rules engines, statistical models, and machine learning algorithms combine to extract deeper meaning from the incoming data. Rules engines apply established operational parameters and domain expertise to detect known conditions. Statistical models monitor patterns and identify anomalies across multiple variables. Machine learning components continue to refine predictions by learning from historical outcomes, helping improve the accuracy of insights over time. These intelligence components operate on different timescales, from millisecond-level safety monitoring to longer-term optimization analyses.

To manage this growing intelligence, the EIH maintains a local repository of models. This includes both site-specific algorithms tailored to local operations and enterprise-standard models designed for broader consistency. Version control ensures that updates align with corporate standards while still preserving the flexibility to adjust to local needs.

The orchestration layer pulls everything together, managing how data flows, when analyses are performed, and how decisions are made. It intelligently determines which actions can be handled locally and which require validation from enterprise systems. Routine process adjustments may be completed entirely at the site level, while proposed changes affecting product quality or compliance are routed for broader review.

Finally, visualization and user interfaces ensure that insights reach the right people in the right way. From control room dashboards to mobile alerts, the Edge Intelligence Hub presents operators, engineers, and managers with information tailored to their roles, supporting faster and more confident decision making.

Deployment of the EIH follows a phased approach. Organizations start by building basic connectivity and monitoring capabilities, then gradually introduce advanced analytics and autonomous decision making tools. This method builds technical strength while helping teams gain trust and familiarity with the system, laying the groundwork for more sophisticated edge intelligence over time.

As these capabilities mature, the role of the Edge Intelligence Hub moves beyond simply collecting raw data for central analysis. Instead, intelligence is distributed directly to the point of need. Data streams from sensors, quality systems, and environmental monitors are continuously analyzed in context, enabling faster and more informed responses.

To better visualize how this system facilitates local needs while maintaining enterprise cohesion, let's explore its integration and data flows within the broader UMDA framework.

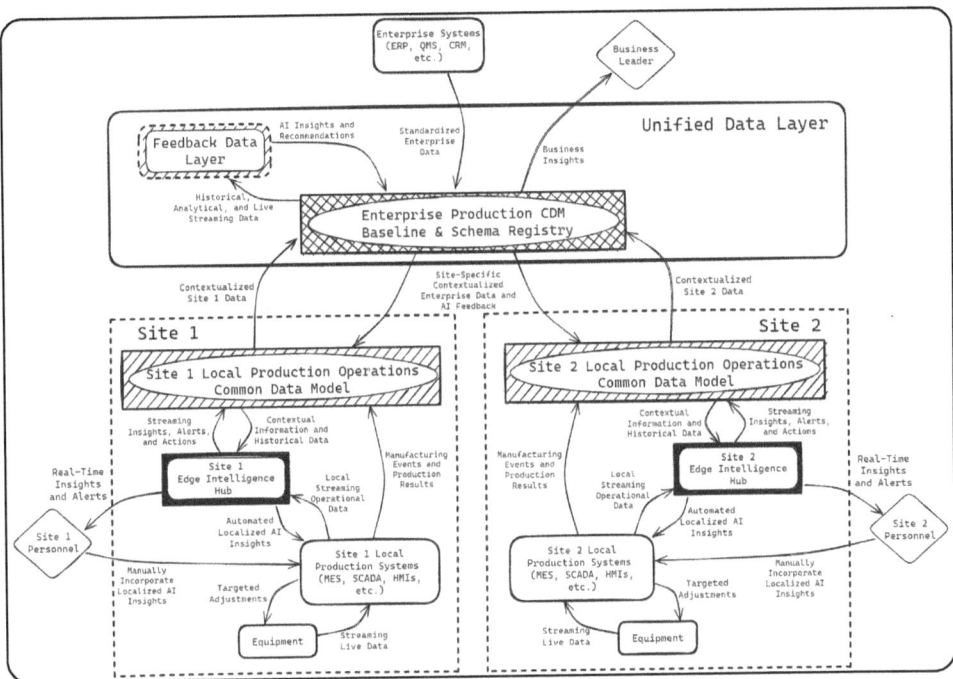

Figure 6-2 Orchestration Between EIH Site Instances and the Enterprise

This diagram highlights how sites work together within the same architecture to balance the need for consistency.

At the enterprise level, the Unified Data Layer serves as the central hub for managing reference data, harmonization rules, and shared identifiers. It provides each site with essential reference materials including quality benchmarks, operational guidelines, and performance metrics. These resources create a consistent foundation, ensuring compliance with standards while supporting alignment across the manufacturing network.

Within the UDL, the Enterprise Production CDM Baseline & Schema

Registry defines a minimal set of shared identifiers and relationships (e.g., process parameters, equipment classes) used to harmonize data across site-level Production CDMs. It does not re-standardize or replace domain CDMs. Instead, it provides mappings and shared reference data to enable cross-site comparability. This distributed structure ensures that core aspects of manufacturing remain comparable and consistent, even as local facilities tailor their operations to fit specific needs.

On the ground at each site, a local Common Data Model bridges the gap between enterprise standards and the unique realities of local operations. The local CDM contextualizes global data, combining it with detailed information specific to the site's equipment, processes, and production goals. This combination enables teams to adapt processes to local needs while remaining aligned.

The real-time decision making layer, represented by the Edge Intelligence Hub, operates directly at the site level. Positioned within the manufacturing environment, it processes live data streams from sensors, quality systems, and operational metrics with minimal delay. This enables immediate responses to changing conditions. By drawing from both local data and enterprise sources, it ensures that every local decision is aligned with the organization.

Another vital component of this architecture is the Feedback Data Layer, which captures and stores the insights, adjustments, and recommendations generated at each site. Whether these come from AI-driven analytics or manual operator inputs, the FDL creates a dynamic repository of learnings. This feedback loop ensures that successful strategies and discoveries at one site can be evaluated, refined, and shared across the organization. Over time, this collective intelligence helps drive continuous improvement and fosters a culture of shared learning.

The benefits of this architecture that incorporate EIH instances becomes evident in practical applications. Consider a scenario where process parameters start to drift, potentially affecting product quality. Traditional systems might issue broad alerts based on static thresholds. In contrast, the EIH evaluates the situation holistically, considering factors like material properties, historical trends, and current environmental conditions. It identifies subtle patterns and recommends targeted adjustments, whether altering machine settings or scheduling maintenance. Operators receive actionable guidance based on current conditions and priorities.

Consider how this plays out in high-pressure scenarios, such as production transitions or sudden equipment failures. By processing real-time data with the added context of site-specific variables, the Edge Intelligence Hub ensures local teams have immediate access to actionable insights. These tailored recommendations help them address disruptions quickly, preventing downtime and maintaining efficiency. The result is a manufacturing environment that thrives even under pressure by adapting fluidly to changing conditions.

The ability of the EIH to deliver real-time guidance proves valuable during high-pressure situations like production transitions or unforeseen events. Operators gain immediate access to a comprehensive assessment of the situation and tailored recommendations, enabling rapid, informed decisions. This speed minimizes downtime, prevents subsequent disruptions, and sustains production efficiency under unexpected conditions.

As manufacturing continues to evolve, the EIH will become a critical engine for innovation, where new ideas are tested, proven, and scaled across the enterprise. This means moving away from waiting for centralized initiatives to drive AI-enabled solutions. Instead, facilities equipped with EIH capabilities become active centers of excellence, solving real-world problems in ways that stay aligned with enterprise standards. By enabling a continuous cycle of development, validation, and sharing, the Edge Intelligence Hub accelerates the growth of local AI solutions while facilitating them to scale consistently across the organization.

Structuring Information Exchange Through Data Contracts

While standardizing data structures is key to scaling AI, it's just as important to remember that data doesn't stand still. It moves between machines, across systems, and throughout the supply chain. It drives predictions, supports decisions, and connects with partners outside the organization. This flow powers modern manufacturing, but without clear rules, it can quickly lead to confusion.

Data contracts help bring order to that movement.

Data contracts define the rules of engagement for information exchange. These are the agreements between the producers and consumers of data that keep everything running smoothly. They go beyond simple data formats, setting expectations for how data should be structured, validated, secured, and shared. When a local plant sends performance data to an enterprise-wide AI model, or a maintenance system communicates with a quality control system, data contracts ensure that every system understands the message the same way.

Without these agreements in place, manufacturing systems risk becoming disconnected and difficult to align. Picture a production line producing thousands of sensor readings every minute, a quality system recording detailed batch data, and a maintenance platform tracking equipment conditions. When each source uses different rules for data structure, timing, or accuracy, the result is a confusing mix of information that doesn't fit together. Even small differences, like one system reporting a temperature in Fahrenheit and another expecting Celsius, can lead to costly mistakes.

Data contracts prevent this disorder by enforcing consistency. They work alongside Common Data Models but serve a different role. While data models provide a standardized vocabulary by defining what each data element means, contracts define how data should be exchanged. They specify the format, the frequency of sharing, who has access, and the data quality standards that must be met. This distinction is critical in a federated manufacturing environment, where local sites need the flexibility to operate independently.

This strategic difference becomes clear when comparing two approaches to data structuring in manufacturing environments. Organizations can either capture raw data first and attempt to apply structure later during analysis, or they can enforce data contracts that align information with Common Data Models right from the source. The choice between these approaches fundamentally shapes how scalable and reliable the resulting data architecture becomes.

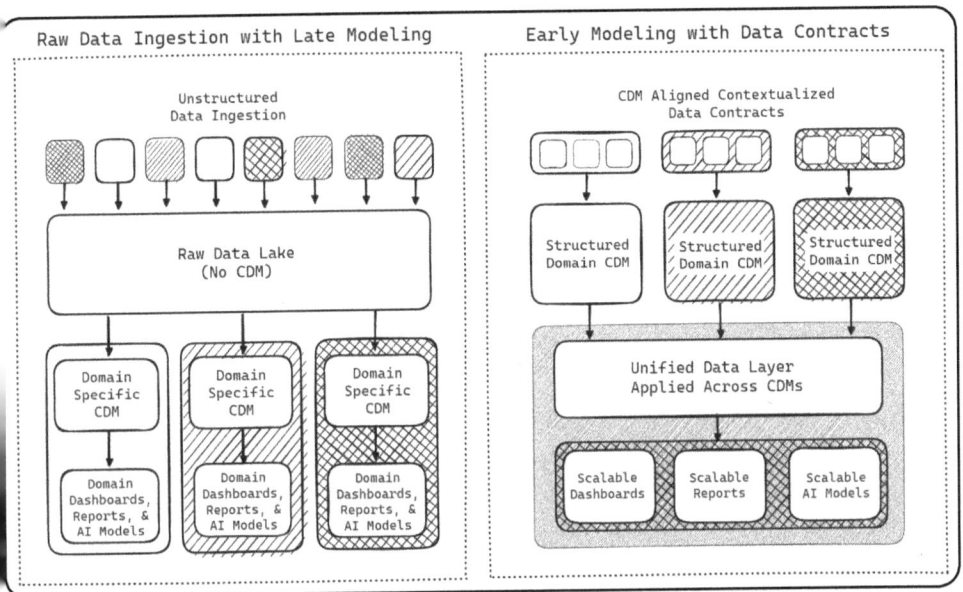

Figure 6-3 Early Data Contract Structure Alignment with CDMs Comparison

The difference between these approaches is dramatic. When data modeling happens late in the process, teams end up building domain-specific solutions that can't easily scale or integrate across domains. Each use case requires custom data preparation, creating new silos that undermine reusable intelligence. But when data contracts enforce CDM alignment at the source, that same consistent data can power dashboards, reports, and AI models in a scalable manner. System owners become accountable for data quality from the moment it's generated, rather than leaving cleanup for downstream consumers.

Within the Unified Manufacturing Data Architecture, data contracts operate at multiple levels. On the factory floor, they ensure that raw sensor data is captured accurately and fed into the Edge Intelligence Hub with the right structure and context. At the site level, data contracts standardize how operational data aligns with the local Common Data Model, making it possible to adapt to local realities while maintaining consistency across facilities. The local CDM acts as the bridge, mapping site-specific data to the enterprise Unified Data Layer. Within the Feedback Data Layer, data contracts structure how

AI-generated insights and human feedback are recorded and shared, creating a closed loop learning system that strengthens both local performance and enterprise-wide intelligence.

The importance of data contracts becomes clear when applied to real-world manufacturing scenarios. A well-defined data contract ensures that every piece of information carries the right context, flows predictably, and supports downstream data consumers. Whether it's optimizing maintenance schedules, aligning production data across multiple sites, or improving collaboration with external partners, these contracts bring order to complex data interactions.

Enabling consistent reporting is another key benefit of data contracts, especially in multi-site operations. Manufacturing plants track work orders, quality checks, and production progress, but without a common structure, comparing performance between sites is nearly impossible. A data contract governing work order updates ensures that every facility reports key events in a consistent way. Whether a site in Germany records a status change due to a machine adjustment or a site in the U.S. flags a quality issue, the update follows the same format, making cross-site analysis seamless regardless of what source systems they were captured in.

External data exchange presents another challenge. Many manufacturers rely on third-party labs for material testing, stability analysis, or compliance validation. Traditionally, this means waiting for finalized reports, delaying decisions and limiting agility. With a data contract in place, external labs can share incremental updates throughout the testing process, following predefined rules for timing, data structure, and validation steps. Instead of waiting weeks for final reports, manufacturers can track trends as they emerge, making proactive adjustments to formulations, supplier relationships, or production schedules.

To see how these concepts work in practice, consider a pharmaceutical manufacturing process like tablet compression. Producing high-quality tablets requires precise control over compression force, weight, thickness, and hardness, with continuous monitoring to ensure consistency. Multiple systems must work together seamlessly including the tablet press, in-process quality testing equipment, the site's manufacturing execution system (MES), and an AI-driven optimization model.

A well-structured data contract ensures the data is exchanged systematically.

The tablet press follows a contract specifying that every 5 seconds, it must report key metrics including compression force, tablet weight, and hardness. It also must send along metadata like batch number, operator ID, and timestamp. This data must be transmitted in a standardized format, with validation rules ensuring values stay within predefined limits.

When the in-process testing system checks tablet quality, another contract governs how results are shared. Each test result is linked back to the exact compression parameters that produced that tablet, maintaining traceability. The contract also specifies response times and requires supporting metadata like calibration status and environmental conditions.

Data contracts don't just standardize information exchange, they create the foundation for a truly intelligent, connected manufacturing ecosystem. With clearly defined rules in place, AI models can operate with confidence, decisions can be made faster, and collaboration across sites and partners becomes seamless. These agreements ensure that every system, whether local or enterprise-wide, speaks the same structured language, making data a reliable asset rather than a fragmented challenge.

Implementing data contracts effectively comes with challenges. Many manufacturers still rely on legacy systems that were not built for modern data sharing needs. A practical solution to integrate these is to create interface layers that serve as a bridge. These interfaces translate legacy data models into the standard formats required by data contracts, supporting compatibility without interrupting current operations. This approach helps organizations modernize gradually, preserving existing infrastructure while enabling consistent and scalable data exchange.

Another challenge is balancing contract rules with operational flexibility. Manufacturing sites have unique processes, systems, and regulatory requirements that require some level of customization. This requires designing modular contracts that establish common standards while allowing for site specific extensions. This way, every facility can tailor data exchanges to its needs without compromising overall consistency.

Organizations that have successfully implemented data contracts start with the highest-value data exchanges, focusing on use cases where inconsistencies or inefficiencies create measurable business impact. Prioritizing these areas builds early momentum, demonstrating value to both technical teams and leadership. Equally important is involving stakeholders from across the organization in defining contract requirements. IT and operations teams need to collaborate to ensure that these agreements are technically sound and practical for real-world production environments.

Once those foundational practices are in place, the focus shifts to enforcement at scale. Modern integration tools make it possible to uphold data contracts across high-volume environments. With so much data flowing through manufacturing systems, manual validation isn't practical. Technologies like API gateways, event driven architectures, and automated validation systems step in to ensure data consistently meets contract specifications in real-time. These tools can flag violations, trigger corrective actions, and even adjust AI model behavior in response to shifts in data quality.

Data contracts play a critical role in maintaining the consistency and reliability of AI models across manufacturing environments. AI systems are only as strong as the data they receive. Inconsistent inputs inevitably lead to unpredictable results. Data contracts set the standards that keep AI models operating predictably across different sites, conditions, and timeframes.

Take the example of a predictive quality model deployed across several production lines. Without data contracts in place, each line might supply slightly different inputs. One might track ambient temperature, while another might skip it entirely. Some might record material viscosity every minute, others only once an hour. These small differences create invisible barriers that weaken AI performance. A model might work flawlessly in one facility but struggle in another, not because the algorithm is wrong, but because the inputs are fundamentally different.

Data contracts prevent these problems by setting clear expectations for both training and operational data. They define which parameters must be included, how often they must be sampled, how to handle missing values, and what validation must occur before the data is used. When an AI model is trained and runs using data governed by these agreements, it stays within the boundaries it was designed for, maintaining reliable performance across locations.

Figure 6-4 Data Contracts form Agreements with Senders and Receivers

They also help tackle another major challenge known as model drift. Over time, as production conditions change, the data patterns an AI model was trained on may drift from current realities, reducing accuracy. By standardizing how data is captured, timestamped, and tied to specific production states, data contracts make it easier to spot these shifts early. Monitoring systems can detect when data patterns start to move, triggering model retraining before performance suffers.

For manufacturers managing multiple AI deployments, data contracts create a solid framework for model governance. They document exactly what inputs a model requires and how those inputs are structured, creating traceability from the raw data all the way to the model's recommendations. This structure is invaluable when troubleshooting, expanding AI into new facilities, or demonstrating compliance with regulatory standards.

As data architectures scale to support emerging Artificial Intelligence solutions, the importance of data contracts will only grow. Without this structure, automated insights can become unreliable or disconnected from real-world operations. With it, AI becomes a trusted partner within a facility and across the enterprise.

Best Practices for Enabling Local Data Mapping

A consistent data framework, supported by clear data contracts, allows manufacturing sites to operate with both independence and alignment. But realizing these benefits requires more than just standardized agreements. It demands thoughtful execution in how local data is mapped to the broader enterprise model. Each site generates unique operational data, and if this information isn't structured correctly, it risks becoming isolated, inconsistent, or unusable at scale. Effective local data mapping ensures that site-specific details are preserved while remaining compatible with enterprise-wide systems, enabling seamless integration, accurate analytics, and reliable AI-driven insights.

At the heart of this process is communication. Local teams need more than a set of guidelines, they need to understand the reasoning behind data standards and how those standards support both their operations and the broader organization. A one-way flow of instructions won't work. Instead, an ongoing dialogue between central data teams and site-level experts ensures that data mapping remains practical, relevant, and adaptable to real-world conditions.

By focusing on collaboration and structured support, organizations can create an environment where localized data mapping drives both flexibility while still remaining scalable. The following best practices help establish this foundation, making it easier for sites to manage their data in ways that benefit both their unique needs and the larger enterprise.

Training plays a major role in bridging the gap between enterprise data strategy and local execution. A thoughtfully designed training program should go beyond one-time onboarding. Instead, it should be a continuous effort that evolves alongside the organization's data needs. Regular workshops, hands-on practice sessions, and opportunities for peer learning help reinforce best practices and ensure that teams stay up to date with new tools and methodologies. Furthermore, this will help to get early buy-in from those who will be executing the mapping as they start to see how their role contributes to the bigger picture.

Not everyone involved in data mapping has the same responsibilities or technical expertise, so training must be tailored to different roles. Data stewards may need deep knowledge of governance policies and data quality management, ensuring they can maintain consistency while allowing for necessary adaptations. Analysts, on the other hand, benefit more from learning how to interpret mapped data effectively, transforming raw values into meaningful insights.

Equally important is creating a system for knowledge sharing. As sites discover better ways to structure and use data, those insights should flow freely across the organization. A central repository of best practices, case studies, and lessons learned can help standardize improvements. Encouraging teams to share their experiences ensures that successful strategies don't stay isolated at a single site but instead contribute to continuous improvement across the entire network.

Ongoing support is just as important as training. Even with a strong foundation, new questions will arise, and challenges will need quick solutions. A well structured support system ensures that local teams can experiment with confidence, knowing they have access to expert guidance when needed. This might include a central help desk staffed by data specialists, a network of subject matter experts available for consultation, and a knowledge base filled with FAQs, tutorials, and best practice guides.

Beyond answering immediate questions, this support system also provides valuable feedback to central teams. If multiple sites encounter the same issues or struggle with certain aspects of data mapping, these insights can drive improvements in training, documentation, or the tools themselves. By treating support as a two-way exchange, organizations refine their processes continuously, making local data mapping more effective across the board.

Collaboration platforms take this even further by enabling real time teamwork across sites. These tools support direct communication between central data architects and local teams, allowing mapping challenges to be addressed as they emerge. For instance, a shared visualization tool might let a process engineer align production data with the Common Data Model while working alongside a remote data architect. This kind of collaboration helps ensure consistency in structure while allowing for the flexibility each site needs.

Shared digital platforms that host mapping templates, workflows, and reusable assets build on this by accelerating the spread of effective solutions.

Instead of each site reinventing the process, teams can adapt proven approaches to fit their local needs while staying aligned with enterprise standards.

Sustaining this coordination requires continuous attention. Data mapping strategies must evolve alongside new technologies, shifting business priorities, and insights gained from daily operations. Treating data management as a living process helps organizations stay flexible without losing alignment. This evolution depends on clear, ongoing communication. Rather than issuing static policy changes, companies should show how new requirements fit into current workflows and actively involve site teams in shaping those changes. When communication is practical and collaborative, it strengthens both local engagement and enterprise-wide cohesion.

For example, when the central governance team introduces new data quality metrics, it takes more than issuing a directive to drive success. Sites need clear guidance on how to integrate these metrics into their processes. They also need the chance to provide feedback on any challenges they encounter. A collaborative rollout helps ensure that new standards are applied, refined, and embedded in daily operations.

Creating a culture of collaboration, continuous learning, and shared accountability strengthens the entire data ecosystem. Organizations can establish forums where data stewards from different sites meet regularly to discuss challenges and solutions. These sessions help troubleshoot common issues while serving as an incubator for new ideas. A process improvement discovered at one facility can be refined, tested, and adopted by others, spreading innovation organically. When teams have access to the right tools and expertise, they can confidently structure and manage data in ways that enhance their operations while contributing to a broader, unified system.

A proactive monitoring framework helps keep local mappings aligned with the Common Data Model as systems and processes evolve. Automated tools can flag discrepancies before they create inconsistencies, while regular audits provide a structured way to evaluate alignment with organizational standards. These mechanisms safeguard the process by ensuring that data remains accurate, accessible, and meaningful at both the local and enterprise levels.

Regular review cycles further reinforce this balance. Scheduled assessments bring central and local teams together to evaluate the effectiveness of current mappings. These sessions provide an opportunity to identify inefficiencies, address evolving requirements, and refine data standards with intention rather than reacting to issues after they arise. They also create a space for sites to contribute insights that may shape broader organizational strategies.

This ongoing refinement process ensures that local autonomy remains valuable. When collaboration, governance, and adaptability become part of the organization's data culture, mapping becomes a strategic advantage. Organizations can then combine local expertise with global oversight, creating a data ecosystem ready to support AI and its contributions to long-term success.

Insights in Action

Norman was halfway through his morning coffee when he noticed something odd on the production floor. Line 2, normally a scattered mix of mechanical jarring noises, was eerily quiet. With a combination of curiosity and concern, he ambled over, his well-worn boots squeaking on the polished concrete.

As he approached, his eyes opened wide in surprise. Line 2 was humming quietly like a purring cat. Norman squinted at the performance dashboard, then let out a low whistle. "Well, look at that," he muttered to himself.

The new sensors they'd installed last month were feeding data into their Edge Intelligence Hub, and the results were astounding. The AI had spotted a pattern they'd never noticed before, suggesting a 1.5% speed adjustment every third hour. This resulted in no more screeching, no more jams, and output was up 12%.

Norman couldn't help but chuckle. He remembered the heated debates about allowing local teams to tinker with data mapping. Now, he was seeing the fruits of that decision firsthand.

Curiosity piqued, he wandered over to the Quality Control station. His eyes widened as he took in the custom dashboard on the screen. "Is that right?" he wondered aloud, squinting at the chart. The defect rate had dropped by 30% since last week thanks to acting on an AI insight from local data points combined with a finding from another site.

As Norman headed back to his office, he felt like he was walking on air.

The factory was humming with efficiency, and he could practically see innovation sparking from every workstation.

Settling into his chair, he opened his email to share the good news. As he typed, he chuckled to himself. "Who would've thought," he mused, "that giving folks a bit of freedom with their data would lead to all this? If this keeps up, we'll be so efficient, I might actually get to use some of that vacation time I've been hoarding!"

With a satisfied grin, Norman hit 'send' on his email. He leaned forward to take a sip of his now-cold coffee, and thought, "Manufacturing sure isn't what it used to be. And thank goodness for that!"

Manufacturing AI

Chapter 7

Building an AI-Ready, Scalable Architecture

||||||||||||||||||||||||||||||||||||||||||||||||||||||||||||||||||||||||||||||||||||

Introduction

Digital manufacturing runs on intelligence. Every sensor reading, quality check, and maintenance log contains information that can improve operations. But extracting that value takes more than just collecting data. It requires an architecture designed specifically for AI that structures and connects information to support real-time decisions, adaptive automation, and continuous learning across the enterprise.

Consider what becomes possible when this architecture is in place. A quality sensor detects the earliest signs of a defect forming while the product is still being made. Within milliseconds, AI systems correlate this data with machine settings, material properties, and environmental conditions. They automatically adjust process parameters to prevent the defect from occurring. Meanwhile, predictive maintenance algorithms spot subtle equipment changes that signal potential failures weeks in advance. Instead of emergency shutdowns, repairs happen during planned downtime.

Companies around the world are deploying AI agents that work continuously, monitoring production lines, optimizing energy usage, and coordinating supply chains. These intelligent systems handle routine monitoring and rapid responses so skilled operators can focus on innovation, problem-solving, and strategic improvements.

Getting to that point brings its own set of challenges. Many legacy systems and older equipment hold valuable data, but they weren't designed for modern AI. Manufacturing teams have to work within the limits of physical processes, where even small changes in temperature or material properties can have a big impact on quality. Older machines, some in service for decades, need to connect with modern data systems without interrupting operations. Many industries have regulations that require every AI recommendation to be clear, explainable, and verifiable.

Organizations that master the integration of legacy and modern sources, respond to market changes in days rather than months. They optimize operations in real-time rather than after problems occur. They scale improvements across facilities rather than repeating projects one site at a time.

Success depends on designing an architecture that serves both immediate and future needs. The system must integrate emerging AI technologies while maintaining the stability and reliability that manufacturing demands. Real-time decisions, security, regulatory compliance, and operational continuity must all work together seamlessly.

This chapter provides your roadmap for building that foundation. You'll discover how to create scalable data architectures that grow with your needs. We'll explore industry standards that ensure consistency across facilities and intelligent routing systems that connect the right data with the right AI models at exactly the right moment. You'll learn to deploy autonomous AI agents that work within defined boundaries, build cross-functional teams that bridge IT and operational technology, and manage risks while maintaining agility.

Through real-world examples and practical guidance, you'll see how companies turn scattered data into coordinated intelligence.

Principles of a Scalable Data Architecture

For AI-driven manufacturing to thrive, data architecture must be designed with scalability in mind. However, this requires more than just accommodating larger datasets or more users. For AI to truly thrive across multiple factories it needs to be able to adapt to new technologies, support evolving workloads, and grow without adding unnecessary complexity. A flexible architecture should seamlessly integrate new data sources, support diverse analytics, and consistently deliver reliable insights, all without the need for constant overhauls.

Achieving adaptability depends on a structured foundation that expands in a way that stays manageable. Within the Unified Manufacturing Data Architecture, this adaptability is realized through a modular, domain-centric approach. Early modeling and standardization occur at the point of data ingestion. Rather than funneling raw data into a central repository for later processing, each data stream is immediately aligned with a Common Data Model. Data contracts define the structure, semantics, and validation criteria as data enters the system, ensuring that inputs from IoT devices, MES platforms, or ERP systems are semantically consistent and ready for enterprise-wide use.

This proactive alignment facilitates scalable data harmonization without relying on fragile, centralized pipelines. For instance, if a facility increases its time-series sensor data volume, the ingestion resources at that site can scale independently, leaving downstream AI processing and analytics dashboards unaffected. Similarly, when a new data consumer such as an AI model is introduced, it can access standardized, context-rich data without needing major changes to the structure of the data sources.

By establishing semantic consistency from the start, the need for late-stage, complex data modeling is eliminated. This approach is particularly beneficial in environments where data originates from multiple factories, systems, and suppliers, each with unique schemas.

Building flexibility starts with clearly defining responsibilities across different parts of the architecture. Data ingestion, semantic modeling, operational context, AI processing, and user interaction are organized into distinct layers.

This structure makes it easier to adapt and grow. Updates to analytics or AI models can happen without interfering with how data is collected or stored. AI retraining can move forward without the need to rebuild entire pipelines, keeping the system both resilient and efficient.

Elasticity is a crucial component of this design. Manufacturing data volumes fluctuate due to factors like product demand, operational hours, and supply chain shifts. The Unified Data Layer must be capable of scaling up during peak activity periods, such as increased production or AI retraining, and scaling down during quieter times. Dynamic resource allocation guarantees consistent performance and cost effectiveness across varying operational conditions.

Importantly, this entire system operates within a domain aware, event driven framework. Instead of depending on periodic system backups or batch data refreshes, the platform captures business events in real-time, each linked to specific operational contexts. Events like production runs, maintenance updates, or quality inspections generate structured messages aligned with the local production Common Data Model. These messages feed directly into the Unified Data Layer, enabling centralized AI, analytics, and operational systems to respond promptly and effectively.

The same principle applies to AI models running at the edge. A predictive maintenance system, for example, may need significant computing power during initial training but far less when running inference on streaming data. A well-designed architecture doesn't just accommodate this shift, it anticipates it, ensuring resources are available where and when they're needed.

To illustrate how these architectural principles come together, imagine a facility producing electronic components with highly seasonal demand. During peak periods production might triple, putting pressure on both equipment and the data infrastructure behind it.

In a traditional system built around a single, tightly coupled platform, this kind of surge could lead to performance issues and system outages. Quality monitoring, predictive maintenance, and production scheduling would all rely on the same shared resources. As the data load increased, the system would slow, forcing trade-offs between real-time monitoring and forward looking predictions. Scaling up to meet demand would mean upgrading the entire platform which is an expensive change that would leave much of the new capacity unused the majority of the time.

Instead of relying on a monolithic system, the manufacturer restructures its architecture around modular and elastic principles. During periods of high production, the value of this modular approach becomes clear. As sensor traffic increases, the ingestion bandwidth expands to handle the load without slowing down storage or processing. AI workloads can shift priorities throughout the day by focusing on real-time quality monitoring during active runs, then turning to maintenance predictions during slower shifts.

Unexpected challenges, like network disruptions, are also easier to manage. Edge processors maintain essential monitoring locally, and alternate pathways keep critical data flowing to the central systems. AI models are built to adapt, continuing to provide guidance even when some inputs are missing. Rather than halting, they adjust thresholds, highlight uncertainty, and support intelligence demands under changing conditions.

The architecture also supports ongoing improvements. A new AI-powered visual inspection system, for example, could be added to the processing layer without reworking other parts of the setup. It would draw from the same shared data foundation, but operate with the specific computing resources it requires. The flexibility to introduce new capabilities without interruption makes expansion more efficient.

Scalable architectures also require ensuring stability under pressure. A system designed to expand must also be able to withstand failures, recover quickly, and continue running smoothly in unpredictable conditions.

Fault tolerance ensures that a system keeps running even when individual components fail. This is achieved through redundancy, failover mechanisms, and error-handling strategies that allow critical processes to continue without disruption. Imagine a data processing node at a manufacturing site suddenly going offline. If the architecture is fault tolerant, backup nodes automatically take over, keeping operations running without data loss or downtime. This kind of built-in redundancy is crucial in manufacturing, where interruptions can lead to production delays, quality issues, and financial losses.

Resilience also requires a focus on rapid recovery and continuous adaptation. A resilient system learns from disruptions, self corrects, and strengthens itself over time. This might involve automated system health checks, AI-driven anomaly detection, and predictive hardware analysis that anticipates failures before they occur.

These principles apply across all layers of a scalable data architecture. At the local level, manufacturing sites need fault-tolerant data pipelines that can maintain real-time processing, even if a server or network link fails. At the enterprise level, the UDL must ensure uninterrupted data exchange across sites, balancing loads dynamically while protecting critical workflows. AI applications also benefit from resilience where models should be able to handle incomplete or delayed data without breaking down.

Common Data Models also need built-in resilience. Synchronizing instances across multiple locations ensures that data consistency is maintained, even if one data model source becomes temporarily unavailable. AI models must be able to function with incomplete or delayed data, adapting to fluctuating conditions without breaking down.

Error handling and self-healing mechanisms further reinforce stability. If a data transformation process fails, the system should attempt automated retries, apply fallback methods, or flag the issue for human review without disrupting the entire pipeline. Manufacturing data environments can't afford cascading failures, making proactive recovery strategies essential. Where possible, the system should prioritize critical functions, ensuring that core manufacturing processes continue even in high stress scenarios..

Leveraging Industry Standards for Consistent Data Models

Industry standards were touched on earlier, but this section will take a closer look at their role in shaping factory-centric data using a standard methodology. Two of the most widely used frameworks in manufacturing are ISA-88 and ISA-95. These standards serve as foundational elements in modern data modeling, providing a structured approach for organizing production activities.

Originally developed for batch manufacturing, ISA-88 has grown into a flexible model used across both process and discrete industries. At its core, it introduces a modular approach to structuring production activities by breaking them into reusable components. This method simplifies complex operations and strengthens automation and AI-driven optimization, making them more effective.

Where ISA-88 focuses on structuring manufacturing processes, ISA-95 bridges the gap between plant operations and enterprise systems. It defines a multi-level hierarchy that organizes data flows between shop-floor control systems and business applications, such as ERP and supply chain management. This structured approach ensures that production activities align with broader business objectives, creating a stream-lined exchange of information between equipment, operators, and management.

Together, these two standards provide the building blocks that manufacturers can use to create reusable, interoperable data models. When AI systems analyze production performance, optimize scheduling, or predict maintenance needs, they rely on consistent data that adheres to these models. Without such a foundation, insights generated at one site may not translate effectively across the organization, limiting the ability to scale AI-driven improvements.

ISA-88 and ISA-95 work together to answer key questions about manufacturing operations, ensuring that processes are neatly structured while also seamlessly integrated into the overall data architecture. To put it simply, ISA-88 focuses on the "why," "how," and "when" of production by defining the purpose, procedures, and timing of every step. While ISA-95 complements it by addressing the "where," "who," and "what." Together creating a complete picture of manufacturing activities.

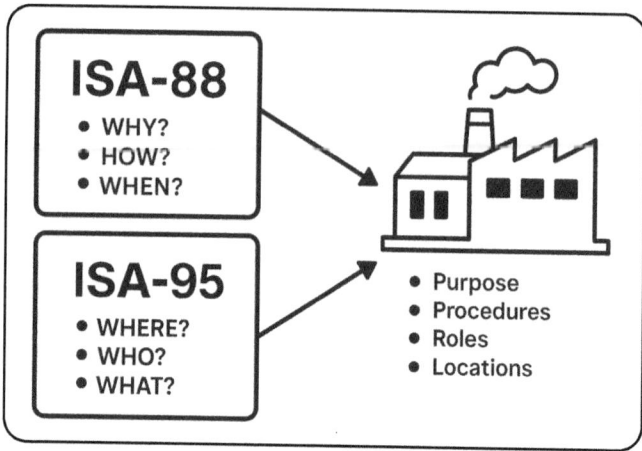

Figure 7-1 ISA-88 & ISA-95 Standards Apply Structure to Factory Data

ISA-88 begins with the "why," defining the purpose behind each step of a manufacturing process. It provides a structured way to represent recipes, process stages, and operational goals, ensuring that production sequences are designed with clear intent. Whether a manufacturer is blending chemicals, baking food products, or assembling complex machinery, ISA-88 helps establish why certain steps must occur, maintaining consistency and repeatability in production.

The "how" is addressed through detailed procedural definitions. ISA-88 breaks manufacturing processes into modular components, such as units, operations, and phases, so that they can be easily reused, modified, and optimized. This modularity allows manufacturers to create flexible, adaptable production processes that can accommodate product variations without requiring a complete redesign. For AI applications, this structured approach makes it easier to automate process adjustments, optimize parameters, and ensure compliance with production standards.

The "when" of ISA-88 dictates the timing and sequencing of manufacturing actions. It ensures that steps happen in the correct order, with precise control over transitions between phases. This is particularly important in industries where timing impacts product quality, such as pharmaceuticals, food processing, and semiconductor manufacturing. With AI-driven process optimization,

ISA-88 provides a framework for making real-time adjustments while ensuring that automated decisions follow established production rules.

ISA-95 builds upon this foundation by starting with the "where" and establishing a clear hierarchy for organizing manufacturing environments from the enterprise level down to individual machines. This structure defines the role of each physical and operational area, supporting consistent data flow and reliable decision making across sites and systems. At the detailed level, it defines Work Units as specific assets like CNC machines or inspection stations, which are grouped into Work Centers representing larger operational areas. This hierarchy provides a consistent way to map where work is done, track performance, and align activities across teams and locations.

The "who" component of ISA-95 defines roles and responsibilities within the manufacturing process. On the shop floor, this clarity supports smarter scheduling, assigning tasks based on skill, certification, and availability. It ensures only properly trained personnel interact with specific equipment or perform regulated procedures. It also helps identify patterns in performance such as highlighting who consistently delivers quality results, who might need support, and who sets the benchmark for best practices. By tying each task and outcome to the people involved, ISA-95 creates a more responsive, accountable environment where AI systems can make context-aware recommendations by including the workforce as a key component of the operational story.

Finally, ISA-95 defines the "what" by standardizing key data exchanges between enterprise systems and the shop floor. It outlines how information such as production orders, material usage, quality checks, and equipment status updates should be structured, ensuring consistent communication across ERP, MES, and AI systems. This standardization enables real time tracking of raw materials, production outputs, and inventory levels. Together this helps forecast more accurately, reduce waste, and keep operations aligned with actual demand.

ISA-95's structured approach to defining manufacturing operations ensures that every asset, role, and material movement is accounted for in a way that supports both local factory needs and enterprise-wide intelligence. By mapping out the physical and logical structure of manufacturing environments, organizations gain a clear understanding of how production activities align with business goals.

The importance of ISA-88 and ISA-95 becomes apparent when considering the specific challenges that artificial intelligence faces in industrial environments. These standards provide a foundational structure for data models required to feed AI, helping the organization align around consistent conventions and reduce debates concerning terminology, structure, or naming.

ISA-88's modular process structure directly addresses several critical AI requirements. When manufacturing processes are broken down into standardized units, operations, and phases, AI models can be trained on dependable data patterns that remain valid across different products and production runs. For example, a machine learning model trained to optimize temperature control during a "heating phase" in one recipe can apply that knowledge to similar heating phases in entirely different products, because the data structure and process definition remain consistent.

This modularity also enhances feature engineering for AI models. Instead of parsing raw, unstructured production data, AI systems can use the semantic information embedded in standardized process definitions. A predictive maintenance model doesn't just see "temperature reading at 2:30 PM", it understands this reading occurred during a specific operation phase, with defined process parameters and expected outcomes. Such context improves model accuracy and reduces training time.

Standardized timing and sequencing offer valuable support for time-series AI applications. Process optimization algorithms can learn what occurs during production and when each step should happen in relation to others. This awareness helps AI detect subtle process drifts, predict quality issues before they arise, and recommend adjustments that preserve efficiency and product specifications.

Hierarchical data organization addresses the challenge of scaling AI insights across different manufacturing environments. Without such a structure, an AI model that optimizes energy consumption at one facility might fail at another due to differing data organizations. Standardizing the definitions of work units, material flows, and personnel roles ensures that AI models trained at one location can be deployed across an entire manufacturing network with minimal adjustments.

The standard's clear definition of roles and responsibilities also enables more intelligent AI decision making. Rather than just creating generic alerts or recommendations, AI systems can understand organizational context. A quality anomaly detected on a production line triggers different responses depending on whether maintenance staff, quality engineers, or production supervisors are best positioned to address the issue. The AI system routes information appropriately because ISA-95 provides a consistent framework for understanding who does what in the manufacturing process.

Crucially, these standards also tackle data quality issues that hinder AI initiatives. In environments with inconsistent data collection practices, AI models spend more time interpreting unclear or contradictory information than providing actionable insights. Establishing clear guidelines for what data to collect, when to capture it, and how it connects to broader manufacturing goals helps filter out distractions, allowing for AI to focus on meaningful patterns.

This clarity in data collection lays the foundation for how AI systems are trained. Once data is consistently captured and well-structured, the next step is to organize it in a way that mirrors the real-world complexity of manufacturing environments. Training data should reflect the hierarchical organization of the manufacturing process by following the ISA standards. Instead of flattening all data into generic tables, it's beneficial to maintain the relationships between different factory levels such as site, area, work center, and work unit, as well as the specific process steps like procedures, operations, and phases.

For instance, organizing datasets around complete process phases (like heating, mixing, or cooling) rather than arbitrary time windows allows AI models to learn the connection between the intent of a process step and its outcomes. This approach helps models understand the purpose behind each action, not just patterns in time-series data.

Feature engineering should also incorporate the context of each data point. Labeling a temperature reading as "heating_phase_setpoint_deviation" provides more meaningful information than a generic "temperature_variance." This semantic labeling enables AI models to recognize patterns that are applicable across different products and processes.

Including material genealogy adds another layer of insight. By linking current process conditions to the history of input materials, AI models can better understand how variations in raw materials affect the final product. This traceability enhances the model's ability to predict quality outcomes and optimize processes.

When designing AI applications, it's advantageous to incorporate the hierarchical nature of the standards into the model themselves. Hierarchical neural networks, for example, can process information at multiple levels simultaneously, learning patterns at the equipment level while considering broader context. Graph neural networks are particularly effective in environments where relationships between work centers, personnel, and material flows are clearly defined.

The combined strength of ISA-88 and ISA-95 goes beyond enabling AI. Together, they form a solid foundation that supports the scalability and long-term adaptability of the UMDA. As manufacturing evolves to include more complex demands like tracking emissions, meeting stricter traceability requirements, or responding to supply chain disruptions, this structure keeps a solid foundation.

When new objectives arise, such as aligning with sustainability goals or complying with new regulatory frameworks, the architecture can be extended without reworking its core. Standards like ISO 9001 for quality and ISO 14001 for environmental management fit naturally into this framework. Their data requirements can be added to existing models in a way that preserves consistent patterns while extending into new domains.

By anchoring data models using these widely recognized standards, manufacturers gain a clear path forward. They can innovate faster, expand into new areas of focus, and maintain alignment across facilities, systems, and teams. This extensibility helps organizations stay flexible without abandoning the structured standards that enable integration and AI effectiveness.

The Unified Namespace Enables Real-Time Data Access

The manufacturing shop floor produces a constant stream of data from PLCs, sensors, machines, and HMIs. While Common Data Models provide the standardized structure that gives this information consistent meaning across the enterprise, there's still the challenge of getting that live data flowing efficiently to where it's needed right now. A Unified Namespace (UNS) solves this real-time distribution challenge by creating the operational data layer that acts as the factory's nervous system, ensuring that consistent data reaches the right systems at exactly the right moment.

The UNS acts as the distribution hub for live operations. While the Unified Data Layer handles persistence, governance, and cross-site intelligence, the UNS delivers the real-time view that keeps everything running in the moment. Every sensor reading, machine status change, and production event flows through this layer at the factory. This makes information available to local dashboards, edge systems, AI models, and analytics tools that need to react immediately.

The Unified Namespace creates an event driven architecture that enables operations through publish-subscribe protocols that deliver real-time updates with minimal latency. When a temperature sensor detects a change, that information doesn't wait in a queue or get processed in batches. It is sent immediately to every system that needs it such as the local HMI, the predictive maintenance model, the quality control dashboard, and the Edge Intelligence Hub that's monitoring that particular process.

The namespace structure follows a clear hierarchy that mirrors operations, often aligned with ISA-95 standards. Every data point gets a meaningful name that instantly reveals its context. A temperature sensor might be labeled as "Site1.BeverageArea4.FinishingLine3.MixingMachine7.TempSensor2", which immediately tells any system exactly where this data originates and what it relates to.

What makes the UNS truly valuable is how it provides plug-and-play components that enable manufacturing systems to grow and adapt. This works because it separates data producers from data consumers. A machine or sensor simply publishes data to the right UNS topic without needing to know who will use it or how many systems are connected. Any application, AI model, dashboard, or analytics tool can subscribe to the data it needs without custom integrations or extra pipelines.

This decoupling creates factory integration flexibility. Teams can add new devices, remove outdated systems, or introduce advanced AI models without reengineering the entire data flow. A new quality inspection camera publishes its data using the established naming convention, and it immediately becomes available to any system subscribed to quality data from that line. The camera doesn't need to know about the predictive maintenance model, the real-time dashboard, or the quality control system that might all be consuming its updates.

The same principle works in reverse. When teams deploy a new AI model for energy optimization, it subscribes to relevant power consumption, temperature, and production rate topics across multiple lines or sites. The model needs no custom connections to each data source. It simply subscribes to required topics through the UNS structure.

AI models tap into this structured, real-time data stream to create live insights. When the Unified Namespace follows Common Data Model standards, that structure carries through to every connected system. A predictive maintenance model built for one mixing machine can easily extend to others because the data looks and behaves the same. There's no need to rewrite the code for different formats or tags. It connects to familiar streams and starts delivering results right away.

This event driven approach becomes particularly powerful for dynamic AI operations. Rather than running models on static datasets, AI can respond to live events as they happen. When a vibration pattern changes, the predictive maintenance model receives that information immediately and can trigger alerts or adjustments before problems escalate. Quality models can spot deviations in real-time and suggest corrective actions while the product is still being manufactured, not hours later during inspection.

The Edge Intelligence Hub uses this data at the source to make quick, informed decisions. It then sends those insights through the Unified Data Layer, connecting local activity with the enterprise perspective.

The following diagram shows a high level overview of how the UNS works as a middle-man between local factory data sources and the Unified Data Layer.

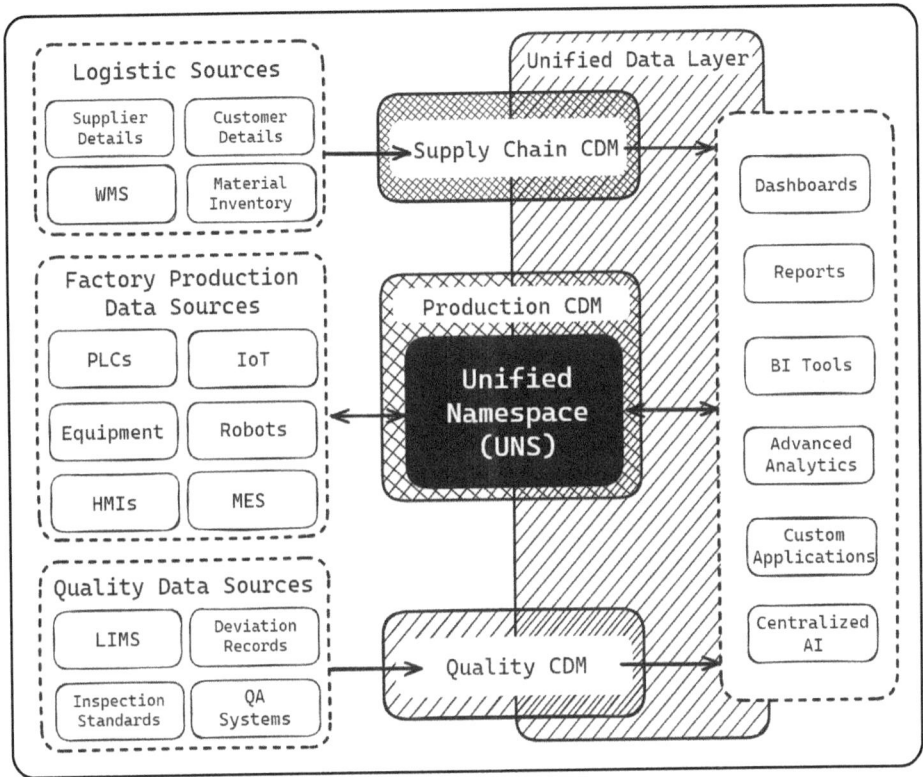

Figure 7-2 The UNS Handles Streaming Production Data at the Site Level

The real-time distribution of information makes troubleshooting much faster. If a packaging line reports a quality issue, maintenance teams can use the UNS to show exactly which equipment is involved, what processes are running, and how materials are moving through each step. The live data stream helps teams trace problems back to their source while issues are still developing, not after they've already impacted production.

The consistency extends beyond individual sites. AI tools monitoring multiple sites can recognize live improvement patterns and evaluate whether similar adjustments could help elsewhere. This immediate feedback loop spreads improvements faster without requiring formal review processes or manual data sharing.

Successful UNS implementations require thoughtful planning. Most effective rollouts start with a focused area of one process, one line, or one set of machines. This gives teams a chance to fine-tune the naming conventions and data flows while demonstrating real value. From there, expansion becomes much easier because the foundation is proven and the benefits are clear.

The structure naturally balances enterprise consistency with site autonomy. Each facility brings its own equipment configurations, process variations, and operational terminology that need to be reflected in the real-time data flow. The UNS accommodates these local differences while maintaining the consistent naming and distribution patterns from the Common Data Model that make cross-site intelligence possible. This flexibility ensures that teams can implement the namespace in ways that match how they actually work, rather than forcing artificial standardization that doesn't reflect operational reality.

Regular collaboration between sites helps maintain this balance as the system evolves. It's important that teams continually share what's working, identify gaps, and coordinate changes that affect multiple sites.

With this real-time UNS foundation in place, the entire architecture becomes more responsive and able to react to streaming factory events.

Instant Insights from Streaming Data

The Unified Namespace sets the foundation for real-time data distribution, but its value comes when that data is processed, analyzed, and turned into insights. Real-time streaming shifts manufacturing from reacting after the fact to staying ahead of issues. It gives teams the ability to spot trends, predict problems, and fine-tune operations while production is underway.

This processing power handles the constant flow of information that the UNS distributes. As consistent data streams arrive from sensors, machines, and operators through the namespace, advanced analytics engines capture it, analyze it against models, and generate insights instantly. If a machine starts drifting from normal operating parameters, the system catches the pattern in real-time and triggers alerts before problems cascade through production.

The complexity lies in processing multiple types of data streams simultaneously at high speed. Structured sensor data, semi-structured logs, and unstructured operator inputs all flow through the UNS and must be analyzed instantly. The processing architecture must scale smoothly as new data sources come online without creating bottlenecks that slow down decision making.

Stream processing makes this real-time analysis possible by working with data in motion rather than waiting for it to be stored and retrieved. This approach is critical in manufacturing environments where delays translate directly into quality problems or production losses. Detecting a temperature anomaly during mixing is far more valuable than discovering it hours later when an entire batch might need to be scrapped.

Complex event processing becomes the key tool for extracting intelligence from these data streams. It analyzes multiple streams simultaneously to identify patterns that single data points might miss. Because the namespace uses consistent naming conventions, streaming analytics can automatically correlate related equipment and processes. When analyzing potential equipment failures, the system can instantly link data from "MixingMachine7.TempSensor2," "MixingMachine7.VibrationSensor1," and "MixingMachine7.PowerMeter5" because the UNS structure reveals their operational relationship.

The Common Data Model context adds another layer of intelligence to stream processing. The same temperature anomaly triggers different responses depending on whether the equipment is processing a high-value pharmaceutical batch or running a standard cleaning cycle.

Real-time responsiveness requires an architecture that can handle everything from simple data filtering to complex machine learning inference on streaming data. Anomaly detection algorithms continuously monitor sensor streams, identifying unusual patterns and flagging potential issues immediately. Predictive models evolve with each new data point, becoming more accurate over time and providing increasingly reliable decision support.

This processing power becomes truly valuable when it connects directly to the systems that control production. When streaming insights flow into Manufacturing Execution Systems and planning tools, they become immediately actionable. Quality outcomes, equipment efficiency, and delivery schedules can all adjust based on current conditions rather than outdated assumptions.

In pharmaceutical batch operations, for example, real-time quality monitoring is essential for both product integrity and regulatory compliance. If streaming analysis detects that a batch is approaching specification limits, AI models can trigger immediate alerts to the MES. The system might adjust process parameters, notify operators, or initiate automated corrections while maintaining the complete traceability that compliance requires. Without this real-time connection, issues might only surface during final lab testing, potentially forcing entire batches to be discarded.

Production scheduling also benefits significantly from streaming data integration. Rather than operating on fixed assumptions, scheduling systems can respond dynamically to actual plant conditions. If a packaging line runs ahead of schedule, the system can advance the next production order. If quality variations reduce yield, it can recalculate material requirements or reallocate resources to maintain delivery commitments.

Live tracking becomes important for maintaining batch genealogy and traceability. It's important to be able to trace each product from raw materials to final output, across multiple production steps and locations. Streaming data keeps these records current as materials move through the process. When quality issues arise, the system can quickly identify the source, determine which batches are affected, and track where those products have been distributed.

Complex manufacturing environments amplify these challenges. A single product might involve dozens of process steps, each generating streams of sensor data, quality measurements, and tracking information. All of these systems must coordinate effectively despite running at different speeds and using different data formats.

Enterprise systems add another layer of complexity. While shop floor data updates every few seconds through the UNS, ERP systems might refresh only several times per day. The streaming architecture must bridge this timing gap by buffering and synchronizing data from fast and slow sources without losing accuracy or creating delays in critical decisions.

When these systems work together effectively, factories can move beyond reactive firefighting to truly predictive manufacturing. By combining live streaming data with historical patterns, AI models can anticipate potential issues or identify optimization opportunities before they impact production. Stream processing powers dashboards that display real-time performance metrics, quality trends, and equipment status, giving teams an up-to-the-second view of operations backed by predictive insights.

This capability proves especially valuable in dynamic manufacturing environments where conditions change rapidly. If ambient temperature shifts affect product quality, stream processing detects the correlation immediately and requests quick adjustments. When sudden demand increases appear, production schedules can be modified in real-time to capture the opportunity. AI can extend this further by recommending or even automating live adjustments like fine-tuning machine parameters or reallocating resources to maintain optimal efficiency.

Edge computing integration makes this real-time intelligence even more powerful. As discussed in the Edge Intelligence Hub section, distributed AI models process critical data locally while participating in the broader streaming architecture. They handle immediate decisions like adjusting machine parameters to maintain quality while sending processed insights to central systems for deeper analysis and enterprise-wide optimization.

This coordination becomes particularly important in time-sensitive processes. Edge devices monitoring equipment for early failure signs can trigger immediate protective shutdowns when critical thresholds are reached. That same data flows through the streaming architecture to assess broader operational impacts and schedule preventive maintenance before problems escalate.

The architecture also supports two-way communication between edge and centralized systems. While edge devices process local data streams, they receive updates including new analytical models and refined AI algorithms that improve performance over time. This creates a continuously learning system that adapts and evolves to meet changing operational requirements.

Advanced AI and machine learning deployed at the edge takes this integration even further. Sophisticated models that once required cloud resources can now run closer to data sources, enabling real-time AI-powered decisions without network dependencies. The combination of edge computing and real-time data streaming represents a fundamental advancement by blending the speed of local processing with layered enterprise intelligence.

Unified Data Across the Enterprise

Creating a consistent data environment for a single line or facility demonstrates what's possible with a well-organized foundation. But the real impact comes when this structure expands across sites, regions, and business units. Companies then realize the benefits of how a unified approach creates broader opportunities for AI and analytics. This growth brings new challenges as each location has different equipment, workflows, and metrics.

Once the initial pilot site is in place, the next steps highlight why consistency is so important. It's easy for each facility to drift into its own version of the data structure. Equipment might be similar, but not identical. Older machines may use outdated formats. Naming habits can reflect years of independent operations. These small differences can make large-scale integration difficult.

To avoid this, successful rollouts start with a shared core structure. This master hierarchy defines the basics and gives teams a common reference point. Local groups then customize parts of it to reflect their specific needs. One site might have extra layers for material flow tracking. Another might include added steps for specialized inspections. These variations keep the data relevant and usable at the local level while preserving a common language that AI systems can work with everywhere.

This phase of growth brings up problems that didn't surface earlier. Older systems may need translation layers to convert their formats. Newer platforms may already be compatible but still require adjustments to fit the broader model. Integrating these systems calls for thoughtful planning and a rollout that leaves room for testing and adjustments.

As the system expands, so does the need for structure and consistency. A single line might manage with basic checks, but when the data multiplies across sites, it's critical to have automated tools in place. These tools help enforce naming rules and detect gaps or inconsistencies as soon as they appear. Regular reviews make sure the structure stays aligned, even as it grows.

The data model must grow, too. As new machines, materials, and business needs emerge, the Common Data Model needs to reflect those changes. While the namespace focuses on how live data is organized, the CDM adds meaning by connecting that data to the broader business picture. This includes product specs, compliance requirements, and operational goals. As companies expand or shift into new product areas, the model needs to be updated to stay relevant.

For example, adding a new product line might bring in new materials, quality checks, or regulatory requirements. These need to fit into the existing structure without breaking it. A business that starts with beverages and moves into food will need to track new factors like allergens, storage conditions, or shelf life.

Different equipment types bring another layer of complexity. Not every facility uses the same machines or automation levels. Some sites might run advanced systems that stream high resolution sensor data. Others might rely on simpler, scheduled maintenance routines. The model needs to make room for both. It should support advanced insights where possible but still deliver value in settings with less detailed data.

As the supply chain stretches across multiple sites, the model must also map those relationships. Shared vendors, linked production schedules, and overlapping material flows need to be visible in the data. This allows for AI tools to track performance across the whole network, not just at one site. Having this bigger picture helps uncover issues like supply inconsistencies, shipping delays, or scheduling conflicts that affect multiple locations at once.

Rolling out AI at scale also reveals new challenges. Even with shared structures, differences in how machines are run or maintained can impact how models perform. What's normal in one facility might trigger an alert in another. These differences can lead to false positives or missed problems if models aren't tuned for each environment.

To manage this, many teams use a layered approach. They might train base models on multi-site data and then fine-tune them locally. Some use learning systems that adjust over time based on local feedback. The goal is to keep insights relevant without building an entirely separate AI system for every site. As more data flows in, these strategies become essential. Local conditions can drift for many reasons such as equipment wear, process adjustments, or even seasonal changes. When this happens at multiple sites, it's easy for models to lose accuracy unless there's a plan in place to keep them updated.

As the architecture evolves, supporting real-time data processing becomes crucial. Traditional batch processing methods may not suffice when AI systems need to make split-second decisions. Implementing low latency data pipelines and ensuring high availability of computing resources are essential steps. Additionally, storage solutions must be optimized to provide rapid access to current data while efficiently managing historical information.

By establishing a robust, scalable data foundation, manufacturers prepare themselves to use advanced AI effectively. With a Unified Namespace organizing factory data consistently and a Common Data Model providing business context, organizations can implement real-time analytics and autonomous AI systems that work reliably at scale.

Handing the Reins to Agentic AI

Artificial intelligence has already delivered meaningful gains in manufacturing, from predicting maintenance needs to spotting defects and managing inventory. These tools have shown their value, but they are limited by waiting for human input, responding to preset conditions, or offering suggestions that still require operator feedback. As data volumes grow and systems become more complex, this reactive model falls short. The next phase requires a shift to AI agents that can take initiative, monitor performance continuously, and make timely decisions without waiting to be asked.

Agentic AI runs like digital independent workers that actively monitor, decide, and respond in real-time. Unlike traditional AI that requires human supervision, agentic AI systems take responsibility for specific tasks, making decisions based on live conditions. Instead of waiting for someone to notice a problem, these systems step in immediately.

AI agents can be compared to specialized factory leaders, each responsible for a different area of the production process. One agent might oversee equipment health, another might manage material flow, and a third could focus on quality control. Each has a defined role and boundaries, ensuring they work within their expertise. As these agents communicate and coordinate actions, their true potential emerges. Every adjustment they make, every outcome they observe, and every challenge they solve improves their ability to respond in the future. Over time, they become highly specialized in their assigned areas, adapting to changing conditions and continuously refining their abilities.

Imagine a digital agent responsible for monitoring critical equipment. It continuously analyzes sensor data, compares it with past performance, and detects early signs of wear or malfunction. But instead of just generating an alert, it evaluates production schedules, checks resource availability, and determines the best time to schedule maintenance. It coordinates with other AI agents managing workflow and material availability, ensuring that repairs happen without disrupting production.

Quality control is another area where agentic AI will have an impact. Instead of relying on traditional inspections, an AI agent can analyze real-time images and sensor data during production. If it detects a drift in quality, it doesn't just flag the issue as a traditional AI model would. It adjusts machine settings, tweaks production parameters, or initiates corrective actions before defects occur.

By implementing agentic AI, manufacturing moves closer to fully autonomous operations. Instead of overwhelming human operators with data and alerts, AI takes responsibility for decision making within carefully defined boundaries. This structure keeps risks low while allowing digital agents to handle tasks intelligently and efficiently.

As AI continues to evolve, handing off these responsibilities to specialized agents is the natural next step. Just like a well-run factory relies on skilled operators for different tasks, the future of manufacturing will rely on AI agents, each focused on a specific role, working together to keep production running at its best.

This flexible approach allows factories to adopt AI at their own pace. Instead of overhauling everything at once, companies can start with the most critical areas. Perhaps assigning an AI agent to monitor a high value machine. Once that system proves effective, additional agents can take on quality control, energy management, and supply chain coordination.

For this to work effectively, AI agents need a strong foundation. The Unified Data Layer acts as a shared knowledge hub, giving each assistant access to real-time sensor data, historical performance trends, and operational context. This ensures that both human operators and AI agents are working with the same facts, making informed decisions based on a full understanding of the manufacturing process.

Giving control to AI to drive actions must follow a structured process. When an AI assistant identifies a potential issue, it doesn't immediately act without oversight. Instead, it logs its recommendation into a feedback system where human operators can review and validate the decision. This creates a built-in safety check, allowing the AI to learn from operator feedback and improve over time.

Take a quality control AI assistant, for example. It continuously monitors production data, analyzing sensor readings and visual inspections in real-time. If it detects a serious defect, it raises an immediate alert or even adjusts machine settings to prevent further issues. For smaller deviations, it logs the data for operators to review later, ensuring that no detail is lost while avoiding unnecessary disruptions.

This creates a continuous cycle of learning and improvement. Machines feed data into AI systems, AI makes adjustments based on that data, and operators provide feedback to refine future decisions. The result is a system that becomes more accurate and reliable with every production run. As repeated findings are discovered, the feedback loop begins to create rules which, over time, enables the AI agents to act on them autonomously.

When AI agents begin making autonomous decisions in manufacturing environments, robust safety and governance frameworks become essential. Unlike other industries where AI mistakes might result in incorrect recommendations or delayed responses, manufacturing environments involve safety-critical systems where poor decisions can lead to equipment damage, product contamination, worker safety, or local environmental impact issues.

Agentic AI systems must integrate seamlessly with existing safety instrumented systems and emergency shutdown procedures. When an AI agent detects a potentially dangerous condition, such as excessive pressure, temperature spikes, or chemical concentration drift, it operates within a hierarchy of safety responses. Critical safety functions remain under the control of certified safety systems, while AI agents handle optimization and efficiency decisions within safe operating parameters. If an agent's recommended action conflicts with safety limits, the safety system always takes precedence, and the conflict is immediately logged for human review.

Regulatory compliance adds another layer of complexity, particularly in industries like pharmaceuticals, food processing, and medical device manufacturing. AI agents operating in these environments must maintain complete audit trails of their decisions, ensuring that every automated action can be traced and justified during regulatory inspections. The system must demonstrate that agent decisions align with validated processes and that any deviations follow established change control procedures. This requirement influences how agents learn and adapt. They cannot simply optimize based on outcomes alone, but must operate within prevalidated parameters that maintain product quality and regulatory compliance.

Agent conflict resolution becomes critical when multiple AI systems have overlapping responsibilities or competing objectives. For example, an energy optimization agent might recommend reducing heating to save costs, while a quality control agent insists on maintaining temperature for product specifications. These systems need clear hierarchies and decision frameworks. Safety and quality will take precedence over efficiency, but the specific resolution depends on business priorities and risk assessments. When conflicts occur, the system can escalate to human oversight while maintaining safe default operations.

Regular governance reviews ensure that AI agents continue operating within intended boundaries as they learn and evolve. Unlike static automation systems, agentic AI adapts over time, which means their behavior can drift from original intentions. Periodic assessments verify that agents remain aligned with business objectives, safety requirements, and regulatory standards. This ongoing oversight helps identify when agents need retraining, boundary adjustments, or additional constraints to maintain optimal performance within acceptable risk levels.

The benefits of this are visible across the entire operation. When a product starts drifting out of specification, the quality assistant can intervene before a defective part is produced. When an equipment monitoring agent detects abnormal vibrations, it can respond within milliseconds, preventing damage and avoiding costly downtime.

This creates intelligent collaboration where AI agents handle routine monitoring and decision making, freeing skilled workers to focus on process improvement, innovation, and strategic planning. AI assistants act as partners, allowing human expertise to focus on creative problem solving rather than reactive alerts.

As AI systems learn, they don't keep that knowledge to themselves. Each assistant contributes its findings to the Feedback Data Layer of the architecture, creating a shared pool of insights across the company. When an energy agent identifies more efficient power settings or a quality control assistant spots how small temperature shifts affect consistency, that information flows to other systems. These insights help other tools adjust and improve, like a climate control system fine-tuning conditions to prevent defects. Even small discoveries can lead to meaningful changes when shared this way, helping the entire operation get smarter over time.

Implementing Agentic AI requires thoughtful planning. AI agents must be designed to work together, communicate with existing systems, and align with specific prioritized goals. Since these systems can be introduced gradually, companies can scale at their own pace, giving employees time to adapt and ensuring a smooth transition.

Setting clear boundaries is essential. While AI agents can operate independently, they must work within defined limits. Each agent should have a clear role, ensuring it stays focused on its task without overstepping into areas where human approval is needed. Regular performance reviews help refine their behavior, ensuring they continue to align with business needs as conditions change.

The future of manufacturing lies in this blend of automation and human expertise. AI-driven systems will enable factories to respond faster, operate more efficiently, and continuously improve without overwhelming workers with data or unnecessary complexity.

Task Management with LLM Routers

As manufacturing systems and operations grow more complex, so do the questions that arise every day. Which machine should be serviced first? Why did quality values suddenly shift? How should production schedules adjust to material shortages? AI agents will help monitor and manage these situations, but they need a way to determine which problems require immediate attention and which need deeper analysis. LLM Routers provide this capability, acting as intelligent coordinators that direct each question to the right system for analysis.

Think of an LLM Router as a highly skilled dispatcher. Just as an experienced supervisor knows who to consult for different issues, the router determines which AI models are best suited to answer specific questions. Some problems require fast, local responses, while others demand broader analysis using data from multiple facilities. The router ensures each request reaches the right model, optimizing both speed and accuracy while managing computational costs effectively.

The routing system operates through three distinct escalation levels, each leveraging different components of the data architecture. At the site level, the Edge Intelligence Hub coordinates lightweight AI models that work directly with real-time sensor data and local process parameters. The EIH provides the local processing infrastructure and decision-making framework that enables these models, built around the Common Data Model structure, to quickly identify obvious issues like sensor malfunctions or process deviations. By managing both the computational resources and the contextual knowledge at the site level, the EIH ensures that most routine problems are resolved locally with immediate responses, eliminating concerns about network delays.

When a problem can't be solved locally, the system routes it to the enterprise level through the Unified Data Layer. Here, advanced models look across multiple sites to spot patterns that wouldn't be visible from just one location. Standardized data structures become essential here. When one system flags an issue with a certain type of equipment, it can quickly check for similar behavior in other locations, without needing custom connections or manual integration work.

For the most complex challenges, the system performs deep analysis using the Feedback Data Layer. This historical repository of insights captures patterns from past resolutions, allowing AI to detect long-term trends and recurring issues. By studying previous outcomes, the system refines its recommendations, ensuring that solutions become increasingly precise over time.

To see how this works in a real-world scenario, let's walk through a typical quality control event and follow how the router manages the decision making process.

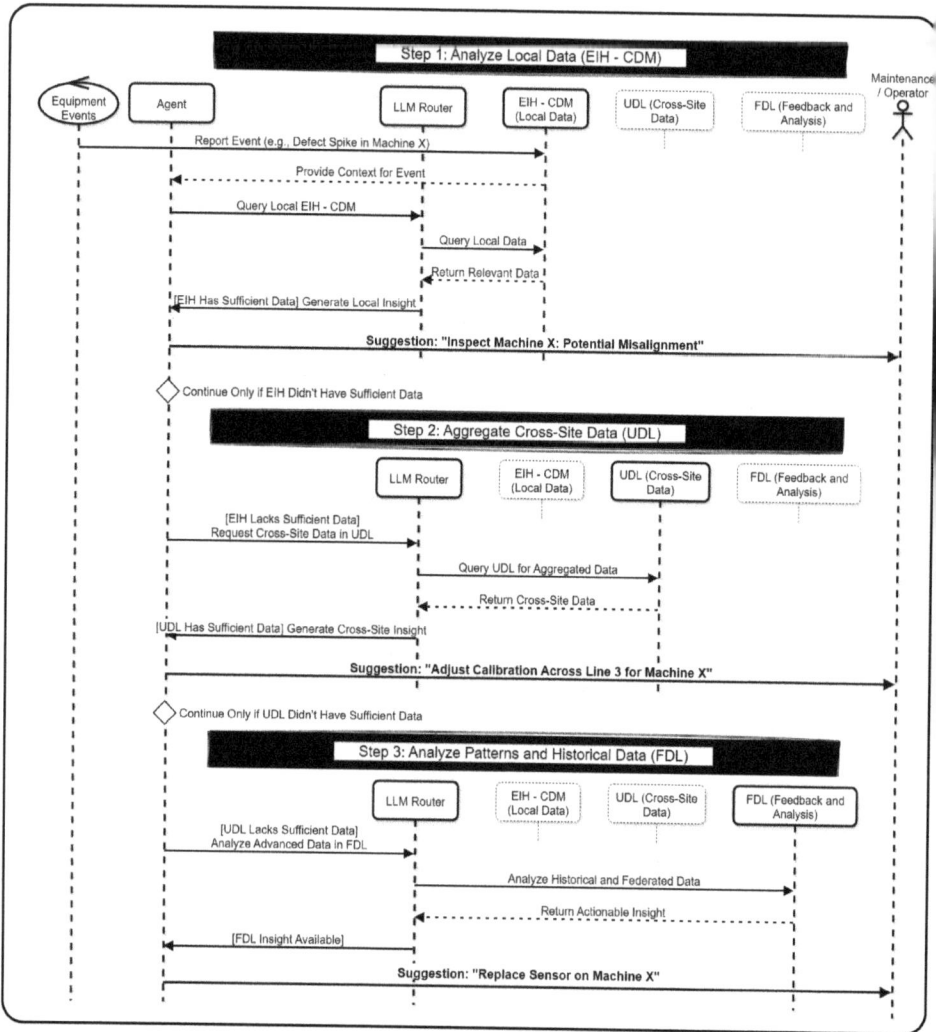

Step 1: Analyze Local Data (EIH - CDM)

Report Event (e.g., Defect Spike in Machine X)
Provide Context for Event
Query Local EIH - CDM
Query Local Data
Return Relevant Data
[EIH Has Sufficient Data] Generate Local Insight
Suggestion: "Inspect Machine X: Potential Misalignment"

Continue Only if EIH Didn't Have Sufficient Data

Step 2: Aggregate Cross-Site Data (UDL)

[EIH Lacks Sufficient Data] Request Cross-Site Data in UDL
Query UDL for Aggregated Data
Return Cross-Site Data
[UDL Has Sufficient Data] Generate Cross-Site Insight
Suggestion: "Adjust Calibration Across Line 3 for Machine X"

Continue Only if UDL Didn't Have Sufficient Data

Step 3: Analyze Patterns and Historical Data (FDL)

[UDL Lacks Sufficient Data] Analyze Advanced Data in FDL
Analyze Historical and Federated Data
Return Actionable Insight
[FDL Insight Available]
Suggestion: "Replace Sensor on Machine X"

Figure 7-3 An AI Agent and LLM Router Coordinate to Diagnose an Event

This escalation process is illustrated through a scenario where an equipment sensor detects a defect spike in Machine X. The agent first is routed to the Edge Intelligence Hub where it analyzes local maintenance records and recent process changes which have been mapped to the local Common Data

Model. This leads to an initial recommendation to inspect the machine for potential misalignment. When local data proves insufficient to resolve the issue completely, the router escalates to the Unified Data Layer, comparing similar patterns across multiple production lines. This broader analysis reveals that the problem affects Line 3 specifically, leading to a recommendation to adjust calibration across all machines in that line. If the issue persists, the final escalation to the Feedback Data Layer analyzes historical maintenance patterns and previous resolutions, ultimately identifying a recurring sensor failure pattern and recommending sensor replacement based on past successful interventions.

To further illustrate how this works in practice, consider a critical component shortage that threatens production schedules across multiple facilities. The disruption triggers immediate analysis through the routing system, demonstrating how different AI models collaborate to solve complex operational challenges.

The process begins when inventory management systems detect that a key component will be depleted earlier than expected. A local AI agent promptly queries lightweight models for immediate alternatives including checking buffer stock, alternative suppliers, and short-term production adjustments. Within seconds, it identifies that two facilities have excess inventory that could cover the shortfall temporarily.

The router identifies the disruption as part of a larger supply chain issue and sends it to the enterprise layer for deeper analysis. Models within the UDL review similar events across other sites, look at supplier trends, and explore alternate sourcing options. The results show a pattern of recurring quality problems from that supplier, something the local team couldn't see on its own. This points to a broader issue impacting many locations.

For the final step, the system pulls from past records stored in the Feedback Data Layer to see how similar supply issues were handled previously. It finds that switching to a different supplier led to better reliability and lower costs. Based on that, the agent recommends more than a quick fix. It suggests making the supplier switch permanent across all sites using that component.

Each stage of this process used the right level of analysis. The router started with a fast, local response that required minimal computing. It turned to deeper enterprise models only when broader patterns needed to be explored.

Historical insights were brought in for strategic decisions, using the most detailed data. This layered method keeps the system efficient while still delivering strong results.

Beyond fixing problems as they happen, LLM routers help shift operations toward continuous improvement. Energy use offers a good example. Local models track power consumption in real-time and adjust equipment schedules during high-cost periods. If something unusual appears, the router sends the data to enterprise systems, which can compare patterns across locations to spot better ways to balance loads or improve equipment efficiency.

Quality control sees the same benefit. Local models catch early signs of process drift and fine-tune settings before defects appear. If a larger trend starts to emerge, enterprise models look across similar lines at other sites to find deeper issues. Once solved, those lessons feed back into the system, improving how future problems are handled across the enterprise.

The system continuously learns from these interactions, refining its routing decisions based on which models prove most effective for specific types of questions. When a particular analysis path consistently produces valuable insights, the router reinforces that pattern. When models provide less useful results, the system adapts, improving accuracy and efficiency over time.

Security considerations are built into every routing decision. Local models retain access to detailed process parameters and proprietary methods, ensuring sensitive manufacturing data never leaves the facility unnecessarily. When broader analysis is required, the system shares only aggregated trends, deviation patterns, or anonymized performance metrics.

This approach extends to regulatory compliance requirements as well. In FDA-regulated environments, for example, the routing system maintains complete audit trails of which models were consulted, what data was accessed, and how recommendations were formed. Critical safety decisions remain under the control of certified safety systems, while AI agents handle optimization within prevalidated parameters.

The system is designed to keep running smoothly, even when preferred models aren't available or produce uncertain results. It uses fallback paths to reroute decisions, so operations continue without interruption. For critical calls, it cross-checks outputs from multiple models to improve accuracy. Users can also see which models were used and why, helping build trust and making oversight easier.

Getting started with LLM routing means picking use cases where smarter decisions lead to clear gains. The best way to build momentum is by starting at the edge with lightweight models that handle common tasks quickly. Once teams see how AI improves daily work, they're more likely to expand its role. Escalation rules guide the process, sending tougher problems or high impact issues up the chain when deeper analysis is needed.

Training AI agents to make smart routing decisions requires ongoing refinement. The Feedback Data Layer plays a crucial role by capturing solutions along with the full context of how problems were resolved. Which analysis paths worked best? What data sources proved most valuable? How did different user roles interact with the system? These insights continuously improve routing logic, making the system more efficient and accurate over time.

Success depends heavily on user adoption and trust. Operators need confidence in AI-driven recommendations and clear understanding of when to rely on automated insights versus seeking additional review. Engineers must know how to frame questions effectively to get meaningful answers. Managers should be comfortable interpreting AI-generated insights for strategic decisions.

The routing system evolves with changing operational needs. New equipment introduces different data types, new products require updated analysis parameters, and supply chain dynamics shift decision making priorities. The system learns from these changes, adapting routing patterns to match new challenges while preserving institutional knowledge about what works best in specific situations.

By ensuring that every manufacturing question reaches the most appropriate analytical resource, these systems help organizations make better use of their data while dramatically improving response times.

The technology enables increasingly sophisticated capabilities as AI models advance and manufacturing intelligence grows. Proactive anomaly detection, dynamic optimization based on real-time conditions, and predictive insights that prevent problems before they occur all become possible when the right analytical resources are applied to the right challenges at the right time.

Cross-Functional Technology Team Strategies

Deploying a Unified Manufacturing Data Architecture requires a coordinated effort between enterprise Information Technology teams and shop floor Operations Technology specialists. These two worlds have historically operated separately, each with its own focus with IT managing systems and infrastructure, and OT overseeing production connectivity and factory centric solutions. But as AI-driven technologies like intelligent agents and LLM routers become central to manufacturing, tighter collaboration is essential.

The solution requires more than occasional meetings between existing teams. Organizations need deliberate strategies for building integrated capabilities that combine enterprise technology expertise with deep manufacturing process knowledge. Success depends on creating new team structures, governance models, and working relationships that treat data as a unified asset rather than separate IT and OT responsibilities.

The most effective way to support intelligent manufacturing systems is by forming unified data teams. Instead of keeping IT and OT separate, these teams combine expertise under one leadership structure. Everyone works toward the same goals. Architects who understand enterprise systems collaborate with automation engineers who know the shop floor. Process experts contribute their knowledge of day-to-day operations, while AI specialists help develop systems that span both technical worlds.

This setup becomes especially important when deploying AI tools that depend on both real-time machine data and enterprise systems. A quality assistant, for example, might need to spot issues using live sensor data and then cross-check that with historical quality records. Executing this task depends on strong OT knowledge to ensure sensors and their production context are accurate, along with IT experience to handle large datasets and keep performance steady.

Supporting this kind of collaboration requires a new approach to governance. Many companies still divide data ownership by system. Machine data stays with OT, business data with IT. But to make this model work, data must be treated as a shared asset. Leading organizations set up governance groups that include both sides. These teams manage data quality, resolve system conflicts, and make sure integration supports both plant operations and corporate goals.

Implementation should follow a phased approach that builds capability incrementally while demonstrating value at each stage. The initial phase typically focuses on establishing basic data connectivity between IT and OT systems, ensuring that information can flow securely in both directions. This infrastructure work requires close collaboration to address different security models, data formats, and performance requirements without disrupting existing operations.

The next phase focuses on building shared analytics capabilities. This typically starts with applications like predictive maintenance or quality monitoring, where combining real-time operational data with historical enterprise information delivers clear value. These early projects help teams gain experience working together and build confidence in integrated approaches. Success comes from choosing use cases where the benefits of cross-domain collaboration are immediately visible to everyone involved.

The final phase introduces more sophisticated capabilities like AI agents that can make autonomous decisions based on data from multiple sources. This advanced functionality requires mature collaboration between teams, robust governance processes, and well-established trust in the integrated approach.

Measuring success in these efforts requires more than tracking IT or OT performance in isolation. Traditional IT metrics focus on uptime, speed, and security. OT looks at output, reliability, and quality. But when teams work together, the metrics should reflect that. The real measure is how well the combined system improves decisions, resolves problems, and supports both plant operations and business goals.

Successful integration also requires a thoughtful approach to change. IT professionals need to deepen their understanding of shop floor realities, including equipment dependencies, production data needs, and the importance of timing. OT teams must become more familiar with cloud tools, data models, and AI systems. Bridging these gaps takes time, but it is essential for creating a truly connected and effective data environment.

Cross-training can help, but real progress happens when teams work together on shared projects. Instead of sitting through lessons on unfamiliar systems, they learn by solving problems side by side. Joint troubleshooting, shared system rollouts, and collaborative planning turn theory into experience. This kind of hands-on work builds stronger skills, better understanding, and the trust needed to support lasting integration.

Leadership plays a crucial role in reinforcing integrated approaches. When executives consistently emphasize that data initiatives must serve both operational efficiency and business objectives, teams naturally look for ways to collaborate. Organizations that succeed at this will generally restructure performance incentives to reward cross-functional collaboration and penalize silo-based optimization that undermines overall system effectiveness.

Many setbacks in this transition come from keeping IT and OT teams separate while expecting them to work closely on shared data projects. Without a common reporting structure and mutual accountability, coordination tends to fall apart when challenges arise. This means aligning incentives and making collaboration part of how teams operate every day.

Another challenge comes from the cultural gap between IT and OT. Each group brings its own language, priorities, and way of solving problems. These differences can lead to confusion and slow progress if they're not addressed early. Teams need time to build trust, learn each other's strengths, and create a shared approach that works across both environments.

The most successful organizations address these challenges proactively by establishing clear communication protocols, shared vocabulary for discussing cross-domain issues, and regular forums for addressing conflicts before they escalate. They also invest in developing internal champions who understand both domains and can facilitate collaboration when teams encounter coordination difficulties.

Technology choices can either support or undermine team integration efforts. Platforms that require deep technical expertise to configure and maintain tend to create new silos, even within integrated teams. The most effective solutions provide intuitive interfaces that allow both IT and OT professionals to contribute their expertise without requiring extensive cross-training in unfamiliar technologies. This accessibility ensures that integration remains focused on leveraging complementary knowledge rather than overcoming tool complexity.

Looking ahead, organizations that successfully integrate IT and OT capabilities build competitive advantages that extend well beyond the immediate technology project. These integrated teams become platforms for rapidly adopting new AI capabilities, responding to changing market conditions, and optimizing operations in ways that pure technology solutions cannot achieve. The collaboration methods, shared governance models, and integrated processes create organizational capabilities that support ongoing digital initiatives across the enterprise.

This shift takes careful planning and long-term focus. But when done right, it opens the door to value that neither IT nor OT could deliver on their own. Treating data as a shared resource, backed by integrated teams and clear governance, gives companies the flexibility and insight needed to adapt and continuously innovate.

Managing AI Risks Through Smart Architecture Design

AI is rapidly creating new possibilities from the factory floor to the extended supply chain. As AI takes on a greater role, it also introduces challenges. Complex production environments leave little room for error. A small miscalculation can spread through the system, affecting quality, safety, and production schedules. Success with artificial intelligence requires more than better models or additional data. It needs a well-structured foundation that keeps AI solutions reliable, secure, and aligned with strategic goals. A strong data architecture ensures AI works within real-world manufacturing constraints by making accurate, trusted recommendations without disrupting workflows.

By structuring data in a way that supports AI-driven processes, the architecture reduces the risks of inaccurate insights, poor decision making, and unintended operational disruptions.

Decades of manufacturing experience have shaped the principles behind this approach. By carefully implementing key components like consistent data models, real-time processing, and governance controls, manufacturers can introduce AI with confidence. This ensures that AI-driven automation enhances operations without compromising reliability.

To fully understand the importance of this structured approach, let's explore some of the key risks AI introduces in manufacturing and how different components of a strategically designed architecture can help mitigate them. From data quality concerns to cybersecurity threats, each challenge requires a thoughtful strategy to ensure AI-driven systems remain accurate, secure, and usable by the workforce.

Data Quality and Reliability Risks

AI can only perform as well as the data it receives. When that data is missing, outdated, or inconsistent, the results suffer. A single faulty sensor, a mismatch in machine setup logs, or conflicting inputs from different systems can throw off predictions and lead to poor decisions. These small errors ripple through production, quality checks, and maintenance routines. Without a clear structure in place to manage and maintain data quality, AI systems are left guessing, and that guesswork can quickly create bigger problems.

Standardizing data across all sources helps reduce these risks. A well-defined data model ensures consistency in how information is structured, eliminating discrepancies between systems. Clear relationships between data points, uniform measurement units, and consistent formats create a shared language for AI and automation, preventing errors caused by misaligned inputs.

Aggregating data while preserving its context is another key factor in maintaining reliability. Instead of treating data points in isolation, manufacturing systems should maintain their connections to operating conditions, process history, and machine performance. AI models that analyze this complete picture gain a more accurate understanding of real-world conditions, reducing the chances of making decisions based on fragmented or misleading data.

Continuous validation and feedback mechanisms further improve accuracy. AI-driven systems should incorporate real-world performance data, operator insights, and historical trends to refine their models over time. When discrepancies arise these systems should detect and adjust for them automatically.

By structuring data properly, maintaining context, and continuously validating AI outputs, organizations can ensure that automated decisions remain trustworthy. A strong foundation of high-quality, well-organized data is essential for reliable AI-driven solutions.

Explainability and AI Bias Challenges

For factory leadership to hand over more decisions to AI-powered solutions, the AI needs to do more than make recommendations. They must provide clear reasoning behind their decisions. When AI suggests adjusting a process, scheduling maintenance, or flagging a quality issue, operators and engineers must understand why. Without transparency, trust in AI weakens, leading to hesitation, resistance, or even the rejection of valuable insights.

Many machine learning models excel at detecting patterns but struggle to explain their conclusions in a way that aligns with human decision making. This "black box" effect becomes a problem when AI suggests actions that don't immediately make sense. If an operator is told to change a machine setting without a clear explanation, they may ignore the recommendation or delay action while seeking clarification. In a fast-moving manufacturing environment, that hesitation can have a huge impact on production.

Bias in AI presents another significant challenge. If historical data contains trends that reflect past inefficiencies or inconsistencies, AI may unknowingly reinforce outdated practices. For example, a predictive maintenance model might recommend servicing a specific machine more frequently simply because past records show more reported failures. But if those records resulted from poor documentation practices rather than actual equipment reliability, the AI could amplify a skewed perception rather than addressing the real problem. These biases can lead to unnecessary downtime, inefficient resource allocation, and missed opportunities for real process improvements.

Addressing these issues requires AI models that are both transparent and accountable. Manufacturing data systems must ensure full traceability, allowing teams to track what data influenced each AI-driven decision. Instead of offering recommendations without context, AI should highlight key variables, historical patterns, and real-time factors that shaped its conclusions.

Causal AI techniques further improve explainability by distinguishing correlation from causation. Unlike traditional models that simply detect associations, causal AI determines the actual drivers of process outcomes. This prevents AI from making misleading assumptions based on surface-level trends and ensures that recommendations are based on true cause-and-effect relationships.

Another key component in managing AI bias and explainability is a structured feedback system where operators and engineers can review AI recommendations, validate their accuracy, and provide direct input on real-world conditions. This oversight helps refine AI models over time, reducing bias and improving trust in automated decision making.

By focusing on explainability, human oversight, and causal reasoning, manufacturers build AI-driven environments where computed insights are clear and trusted. When AI decisions are transparent and grounded in real data, operators and engineers feel confident applying them to improve efficiency and quality.

Safeguarding AI Decisions with Automated Reasoning

Significant risks can be introduced when systems make decisions without proper oversight. AI models excel at detecting patterns and optimizing processes, yet they lack an inherent understanding of physical constraints, safety standards, and regulatory requirements. Without safeguards, an AI model might suggest process adjustments that improve efficiency on paper but create real-world problems such as pushing equipment beyond safe operating limits or misidentifying quality defects.

Automated reasoning provides a critical layer of verification, acting as an intelligent checkpoint before AI-driven recommendations are implemented. Unlike traditional rule-based systems that check for predefined conditions, automated reasoning uses logical inference to evaluate decisions against a broad set of real-world constraints. It considers equipment capabilities, safety thresholds, regulatory compliance, and operational best practices to ensure AI recommendations align with both engineering logic and business objectives.

When an AI system suggests changes to a manufacturing process, automated reasoning first verifies whether those changes are physically possible. Can the equipment actually operate at the suggested settings? Are the proposed adjustments within safe operating ranges? Will the changes maintain product specifications? These checks prevent AI from making decisions that might work in theory but fail in practice.

Automated reasoning also plays a key role in maintaining process stability. AI systems may optimize a single variable for efficiency, but manufacturing processes involve interconnected systems where one adjustment can create unintended ripple effects. By evaluating AI-driven recommendations within the broader context of the entire operation, automated reasoning ensures that changes don't inadvertently create a safety risk, disrupt production, cause quality issues, or introduce long-term risks.

Beyond immediate verification, automated reasoning improves over time by learning from past decisions. When AI recommendations are approved or rejected, the reasoning system captures those outcomes to refine its validation criteria. If a proposed adjustment repeatedly fails verification due to safety concerns, the system learns to flag similar recommendations earlier in the process. This continuous feedback loop strengthens AI oversight and reduces the likelihood of errors.

In complex, multi-step processes, automated reasoning evaluates entire sequences of AI-driven decisions. Instead of verifying individual changes in isolation, it assesses whether a chain of adjustments will maintain operational stability throughout. This approach prevents AI from making changes that, while seemingly acceptable in the short-term, could lead to process imbalances or cumulative risks over time.

By integrating automated reasoning into manufacturing workflows, companies can confidently scale AI solutions while maintaining control over operational safety, compliance, and efficiency. This added layer of intelligence ensures that AI remains a reliable tool for process optimization rather than a source of unpredictable risk.

Cybersecurity and Data Privacy Risks

As legacy controllers and systems become more connected to support data needs, the attack surface expands. Cybersecurity and data privacy become more critical than ever. AI systems pull data from equipment, production line control systems, enterprise platforms, and external partners, creating more entry points for cyber threats. What was once a closed system is now an interconnected network, blending traditional IT infrastructure with operational technology that controls physical processes. This convergence brings powerful capabilities but also introduces new risks.

A security breach in a manufacturing AI system can disrupt production, manipulate quality results, or alter critical machine settings. Attackers who gain access to AI-driven process controls could cause real harm, from shutting down operations to physical disasters. Furthermore, protecting proprietary manufacturing data, including process parameters, quality standards, and optimization strategies, is essential for maintaining a competitive edge. Intellectual property theft or unauthorized access to supplier and customer information can have lasting consequences.

A strong security strategy starts with limiting access to critical data. Clear access controls ensure that only authorized users and systems can retrieve or modify information. Encrypting data, both at rest and in transit, prevents unauthorized interception or manipulation. Authentication measures, such as multi-factor authentication and role-based permissions, help verify every system and person accessing AI-driven insights.

Processing data closer to its source can also reduce security risks. Instead of transmitting sensitive machine data across networks, edge computing allows AI models to analyze information locally, limiting exposure to cyber threats. This approach also improves response times, ensuring that real-time decisions remain secure and uninterrupted.

AI itself can be leveraged to enhance cybersecurity. By continuously monitoring network activity, equipment behavior, and data flows, AI-powered anomaly detection systems can identify irregular patterns. If unexpected access attempts, data manipulation, or deviations in process controls are detected, the system can trigger immediate alerts or automated responses to contain potential threats before they escalate.

By embedding security into every layer of operations, manufacturers can minimize risks while maintaining the agility and efficiency that these technologies provide. As AI adoption grows, securing data, controlling access, and using intelligent monitoring will be key to ensuring a safe and resilient manufacturing environment.

Handling Cultural Resistance

Manufacturing has long valued hands-on experience, reliable processes, and direct interaction with equipment. This practical mindset doesn't always align with the rollout of AI solutions. When intelligent systems suggest changes based on data patterns or algorithms, it can feel disconnected from what operators see and trust on the shop floor. If those recommendations go against standard practices or long-held instincts, it's natural for teams to question them. This is especially true if AI efforts have previously failed at the site.

Trust becomes a central challenge when introducing AI into environments where decisions directly impact safety, quality, and production targets. Manufacturing workers understand that small errors can grow into significant problems, making them appropriately cautious about recommendations from systems they don't fully understand. If AI suggestions lead to quality issues, equipment problems, or safety concerns, resistance solidifies and becomes much harder to overcome through subsequent initiatives.

Building acceptance starts with proving AI's value through early applications where the benefits are clear and the risks are low. When AI predicts equipment failures that might have gone unnoticed, it earns trust without disrupting established ways of working. The same applies to quality monitoring, where AI helps spot issues sooner while keeping operators in control of how to respond.

Transparency plays a crucial role in reducing resistance. Workers need to understand what AI systems recommend and why those recommendations make sense within their operational context. This requires AI implementations that provide clear explanations connecting guidance to observable conditions, historical patterns, and established engineering principles. When operators can trace AI logic back to factors they recognize and trust, acceptance increases significantly.

Successful change management also requires acknowledging legitimate concerns about job security and role changes. Rather than dismissing these fears, effective leaders address them directly by demonstrating how AI enhances rather than eliminates human capabilities. This might involve redefining roles to emphasize higher-level problem solving, expanding responsibilities to include AI system oversight, or creating advancement opportunities that combine traditional manufacturing skills with AI expertise.

Bringing workers into the process early makes a difference too. When operators help shape the system by contributing data, testing early versions, or giving feedback, they're more likely to support the final product. It becomes something they helped build, not something imposed on them. That sense of ownership, combined with a steady rollout and clear communication, makes change feel less like a disruption and more like a new tool that enables progress.

Skills Gap and Workforce Displacement Concerns

As experienced operators and engineers near retirement and AI adoption accelerates, manufacturing faces a critical workforce transition. This timing creates a dual challenge of building new digital skills while preserving decades of hard-earned knowledge before it disappears.

Senior technicians bring deep, intuitive knowledge of equipment behavior, process anomalies, and optimization strategies that are hard to capture through documentation or formal training. Most of these seasoned technicians, however, have limited experience with interpreting and validating AI outputs. Newer workers may be more comfortable with digital tools but may lack the process expertise needed to oversee AI-driven recommendations effectively. This creates a dual challenge of preserving the wisdom of experienced workers while strengthening AI literacy across the workforce.

A strategically designed data framework helps close this gap by preserving institutional wisdom. The Feedback Data Layer is a primary contributor to this. It logs results, measurements, and the reasoning behind successful actions along with the thought process that led to decisions. This archive of experience helps both AI systems learn and new workers understand the logic behind past choices.

Effective talent development strategies recognize that AI should enhance traditional expertise. Workers learn how to interpret AI insights within their own process knowledge. Engineers learn how to check AI-driven recommendations against real-world constraints and safety expectations.

Mentorship is important in this journey. Pairing seasoned operators with newer staff creates opportunities for hands-on learning. When senior team members use AI recommendations as teaching moments, they pass on must-have knowledge while building trust in new tools. This collaborative model reduces fears of job displacement and positions AI as a tool that enhances human skill.

Companies that successfully navigate this transition define new career paths that blend technical and manufacturing expertise. For example, a technician may become an expert in AI-powered predictive maintenance, or a quality engineer may supplement their traditional skills with digital optimization tools. These hybrid roles show that AI implementation can drive career advancement rather than cut jobs, building a stronger, more capable workforce for the future.

Compliance and Ethical Considerations

Manufacturers operate under strict regulatory standards that differ by industry. Whether it's FDA rules for pharmaceuticals or safety standards for automotive systems, these frameworks set clear expectations for how decisions must be made and documented. As AI begins to influence quality checks, safety controls, and environmental monitoring, those systems must prove they meet regulatory requirements. The challenge is that many of these rules were written before AI came into play, leaving room for confusion about what counts as compliant when automated algorithms are involved.

Regulators now expect manufacturers to show that AI decisions are grounded in sound engineering, not just statistical patterns. In pharmaceutical production, for example, a system that flags quality issues must also explain why. Inspectors need to verify those decisions against established standards. That's difficult when many machine learning models operate as black boxes. Organizations must find the right balance between model performance and decision transparency.

Ethical concerns go hand-in-hand with compliance. When AI recommends changes that could affect safety or quality, there needs to be a clear protocol for human oversight. This gets more complicated when decisions happen faster than people can react. That's why systems need built-in safeguards that keep AI within safe limits, even when operating on its own.

Bias from historical patterns adds another layer of risk. AI trained on past performance data might reinforce old inefficiencies or overlook new safety practices. Worse, if those systems are used in workforce management or performance tracking, they could inadvertently discriminate against certain employee groups if historical data reflects biased management decisions or unequal working conditions. This can impact employee safety, fairness, and trust.

To manage all of this, companies need strong governance. That means setting clear rules for how AI makes decisions, ensuring every action is traceable, and validating recommendations against both compliance standards and internal values. These controls should be part of the broader quality and compliance systems already in place. When AI governance is integrated into existing quality management and regulatory compliance processes, organizations can scale AI adoption while maintaining the trust and accountability that manufacturing operations require.

ROI Failure and Business Case Risks

Most manufacturing companies find it difficult to measure the return on investment from AI projects. Unlike traditional physical automation, where success may be easy to track, AI tends to improve many areas a little at a time. It can lower maintenance costs, reduce waste, improve quality, and help managers make informed timely decisions. But because these gains are distributed across departments, they're harder to quantify. This makes it challenging to show executives where the value is, especially when upfront costs are visible but the benefits take time to emerge.

The nature of factory operations adds to the challenge. AI solutions can evolve to support more than the original task it was designed for. A system built to prevent machine failures might be extended to improve scheduling, reduce last-minute jobs, and boost customer satisfaction. These ripple effects matter, but they aren't always captured in traditional ROI accounting. This can make the true return appear smaller than it really is.

Problems also arise when teams aim too high too early. It's tempting to tackle the toughest problems first, thinking the payoff will be worth it. But the most complex use cases involve more data, more integration, and more time. These projects are also more likely to run into delays. When they do, they can shake confidence in the entire AI platform rollout.

A better way to build the business case starts with knowing where things stand. Before launching a project, teams need to understand current costs, performance, and bottlenecks. This process can highlight simpler fixes that should be made before AI gets involved. Fixing any operational issues before starting helps ensure that AI builds on a strong foundation, rather than trying to cover for weak processes.

Setting realistic timelines is just as important. AI doesn't deliver value the moment it's turned on. Good models need time to learn. Operators need time to build trust and figure out how best to use the insights. If a business case promises instant results, it sets expectations that are hard to meet and can overshadow the eventual long-term value.

A well-designed data architecture helps reduce these risks. When the right structures are in place, AI projects can get up and running faster and with less effort. This lowers the cost and shortens the time it takes to see results. It also allows teams to start small, prove value, and build momentum through experience. By scaling gradually, teams learn what works while building up to more impactful business cases as they go.

Managing AI Risk - Conclusion

Successfully integrating AI into manufacturing demands a structured approach to risk management, governance, and big-picture system design. A strong data foundation, well-defined oversight mechanisms, and logical guardrails make AI-driven manufacturing both scalable and reliable.

Balancing automation with human expertise is key to preventing unintended consequences. AI can help optimize production, but without proper oversight it can also amplify errors or make decisions based on bad or incomplete data. Clear boundaries and structured feedback loops keep AI decisions aligned with operational goals while allowing human judgment to intervene when needed.

The quality, consistency, and security of data play a central role in reducing risk. Ensuring data is well structured, properly contextualized, and free from bias reduces the likelihood of AI decisions negatively impacting production processes. Furthermore, strong cybersecurity measures protect critical information from manipulation, unauthorized access, and external threats, safeguarding AI systems from potential disruptions.

Automated reasoning further strengthens AI decision making by applying logical verification before recommendations are implemented. Rather than blindly acting on patterns detected in historical data, AI systems can cross-check their decisions against safety constraints, past performance, and operational requirements. This extra layer of validation helps prevent costly mistakes and ensures AI remains an asset rather than a liability.

As AI capabilities expand, its role in manufacturing will continue to grow. But with each advancement also brings new challenges. A thoughtfully designed architecture provides the structure needed to scale AI solutions with reduced risk, ensuring that emerging technology remains a tool for progress.

Insights in Action

Norman sat at his desk, his eyes fixed on the large monitor displaying a unified dashboard of the factory's operations. A production slowdown had been flagged, and in the past, this would have meant hours of digging through disconnected systems to find the root cause.

He mused, remembering the frustration of those days. "This used to be like looking for a needle in a haystack," he muttered. "ERP data in one system, MES in another, and the shop floor Historian speaking a language all its own."

But today was different. The new connected architecture that bridged the IT and OT systems was about to prove its worth.

As Norman watched, real-time data from the production line flowed into the system, automatically being contextualized with information from the ERP and Quality system. A sophisticated AI algorithm, designed to detect anomalies and correlations, was processing this unified data stream in real-time.

Suddenly, an alert popped up on the screen. The AI had detected a pattern.

"Well, would you look at that," Norman said, leaning forward with interest.

The system had correlated a recent change in raw material supplier, logged in the ERP system, with subtle variations in machine performance on the production line. These variations weren't significant enough to trigger any individual alarms, but the AI, with its holistic view of the data, had spotted the connection.

Norman quickly pulled up the material specifications. Sure enough, the new supplier's materials were within accepted tolerances, but at the lower end of the range. This slight difference was causing minor but cumulative issues in production, leading to the overall slowdown.

With a few clicks, Norman sent alerts to the procurement team about the material issue and to the line managers with instructions for fine-tuning the machines to accommodate the slightly different raw material properties.

He sat back, shaking his head in amazement. From problem detection to root cause analysis and solution implementation, all in a matter of minutes. The new unified IT/OT architecture, with its real-time data streaming and AI-powered analytics, had turned what used to be a time-consuming, frustrating ordeal into a smooth, efficient process.

Norman mused, "Who would've thought that connecting all these systems would be like giving the factory a brain of its own? It's now making immediate connections we might never have found."

As he watched the production rates begin to normalize on his screen, Norman felt a sense of pride. The factory was entering a new era, and despite his initial skepticism, he was excited to be part of it. With this new unified system, they weren't just reacting to issues anymore, they were staying one step ahead.

Chapter 8

Future Trends in Manufacturing Technology

||||||||||||||||||||||||||||||||||||||||||||||||||||||||||||||||||||||||||||||||||||

Introduction

We are now entering an era where technology will completely redefine how factories operate. The factories of the future will be fluid, autonomous, and capable of adapting to changes in ways that extend far beyond today's shop floors. AI won't just optimize processes, it will redesign them. Machines won't just follow programmed instructions, they will reason, predict, and collaborate. Deep analytics won't just support decisions, it will drive entire systems in real-time.

What happens when AI moves beyond recommendations and takes on more autonomous roles? As AI agents gain the ability to plan, coordinate, and act across entire production environments, manufacturing will shift from human led decision making to a model where AI and automation continuously self optimize. The traditional concept of the production line may disappear altogether, replaced by modular manufacturing cells that reconfigure themselves on demand.

Self-optimizing systems will change how products themselves are designed and made. AI will help develop new materials with properties tuned for strength, flexibility, or sustainability, right down to the molecular level. That means bioengineered surfaces that repair themselves or coatings that adapt to changing environmental conditions could move from research labs into everyday use. As manufacturing evolves, the traditional model of producing separate parts for later assembly may give way to continuous processes. Technologies like additive manufacturing and programmable materials will allow complete products to take shape in one seamless step.

As quantum computing matures, it will eclipse existing limitations in manufacturing simulations, supply chain optimization, and materials science. Digital twins will evolve from monitoring tools into interactive AI-powered environments. This will give manufacturers the ability to test and improve processes long before anything happens on the production floor.

While AI will have a big impact on how things are made, it also will influence where they're made. Manufacturing's relationship with supply chains will evolve as AI-driven ecosystems predict and prevent disruptions before they happen. Distributed manufacturing networks will allow production to shift seamlessly between locations based on demand, resource availability, and geopolitical factors.

This shift in where things are made is just as important as how they're made. Instead of shipping products across continents, decentralized micro-factories will produce goods closer to the end user, reducing costs, waste, and delays. These microfactories could dynamically switch between product types based on real-time demand such as manufacturing electronics in the morning, personal mobility devices in the afternoon, and aerospace components overnight, all reconfiguring on the fly with no downtime.

This shift will open the door for more manufacturers to compete at a higher level. AI will give smaller operations access to tools that were once only available to the largest players. On-demand distributed cloud platforms will offer real-time optimization, simulation, and automation without the need for deep in-house expertise. Instead of building their own infrastructure, companies will tap into AI solutions as a service, ready when and where they need them.

The workforce will change significantly. AI will handle many routine tasks while creating new roles that combine engineering, data science, and systems thinking. Jobs will shift toward managing AI ecosystems rather than individual tasks. Career paths will become more flexible, evolving with technical advances. Human creativity and problem solving will matter more than ever, as people partner with AI systems in daily operations.

Sustainability will be embedded in every process. AI will optimize material usage by designing components that generate less waste while maximizing durability. Factories will generate much of their own energy, capturing and reusing excess heat, kinetic energy, and even carbon emissions to fuel production. Recycling and circular manufacturing models will be driven by AI decision making, ensuring that materials are reused efficiently with minimal environmental impact.

Success in this future depends on designing systems that can grow and adapt as technology evolves. The architectures built today need the flexibility to connect with AI agents that have yet to be developed, machine learning models that refine themselves over time, and quantum computing tools that will push past current computational limits. Organizations that build for adaptability will move ahead, while those tied to rigid models will find it hard to keep pace.

The pieces of the next technology wave are already coming together, though its full impact is still to come. This chapter looks at what lies beyond the systems covered so far. It examines how AI will take on a greater role in shaping manufacturing itself, how autonomous systems will redefine factory operations, and how businesses can set themselves up to lead in this changing landscape.

The next industrial revolution is already underway. The question is, who will shape it?

The Rise of AI Predictive Analytics

We are approaching a point where AI predictive analytics will drive most major decisions. Instead of waiting for issues to appear and reacting to problems as they unfold, manufacturing systems will anticipate disruptions, optimize solutions in real-time, and take proactive steps without human intervention. This shift will create algorithms that predict failures, adjust to market shifts, and fine-tune production efficiency without ever needing to pause.

Future systems won't only detect patterns in machine performance and supply chain logistics but will also understand cause-and-effect relationships that people might miss. These intelligent networks will see connections across thousands of variables, from subtle vibrations in machinery to shifts in global commodity prices, creating a level of foresight that advances reactive operating methods into predictive practices.

Predictive analytics is already changing how production lines run in many modern factories. These systems do more than forecast when something might break. They catch small changes, such as shifts in performance that come before a problem starts. If humidity changes affect material properties, AI can adjust settings on the spot to keep quality consistent. If a machine starts to drift from its normal range, the system doesn't wait for it to fail. It adjusts workloads across the plant to avoid downtime. These real-time responses keep operations running smoothly while improving reliability across the board.

AI is changing the way quality is managed on the production floor. Instead of spotting defects after the fact, smart systems now prevent them from happening in the first place. They track thousands of variables such as raw material conditions, machine settings, and past defects. With that information they adjust processes in real-time to keep everything running within spec. That means fewer surprises, less waste, and fewer recalls. Quality becomes something built into the process, not checked after the fact.

The same thinking now applies to supply chains. Rather than reacting to problems, teams can spot them before they hit. Predictive systems monitor supplier performance, shipping activity, and global events to flag risks early. If a shipping port shuts down or a supplier's schedule starts to slip, the system can shift orders, reroute materials, or change production schedules to avoid delays.

That kind of foresight keeps everything moving, no matter what's happening across the world.

Product design is being redefined by AI's predictive capabilities. Instead of engineers manually iterating on prototypes, AI-powered design systems simulate thousands of variations in minutes, predicting performance outcomes before physical models are built. By incorporating real-world production constraints into these simulations, AI creates designs that are both innovative and optimized for manufacturability and cost effectiveness. A new automotive component might be designed for strength and weight, while also considering how efficiently robots can assemble it and how likely it is to pass quality inspections on the first run.

AI becomes more powerful with every production cycle. Over time, it refines its understanding of how materials behave, how machines age, and how conditions change. What starts as a system that needs human validation evolves into one that handles adjustments on its own. It can fine-tune settings when materials shift, adapt to equipment wear, and keep production steady without operator input. This evolution represents a fundamental shift from human intuition to automated pattern recognition across massive datasets.

The most advanced predictive systems will operate by coordinating every aspect of manufacturing in real-time. When a supplier delay is detected, the system does more than simply alert managers. It evaluates alternative suppliers, calculates the impact on production schedules, adjusts inventory levels, and modifies product priorities to minimize disruption.

Machine learning models will evolve to predict cascading effects of outages across the entire value chain. If a critical machine shows early signs of wear, AI will calculate how that affects downstream processes, supplier schedules, and delivery commitments. It might recommend shifting production to alternative lines, ordering replacement parts before they're needed, or adjusting customer delivery schedules to prevent service disruptions.

The convergence of IoT sensors, edge computing, and advanced AI is creating manufacturing environments that learn and improve autonomously. Every temperature reading, vibration measurement, and production metric feeds into models that become smarter with each data point. These systems will identify optimization opportunities that people would never notice, like discovering that a slight adjustment in one process step improves quality three stations downstream.

Manufacturers are no longer limited to reacting to what's already happened. With predictive analytics guiding decisions, factory operations become more agile, more resilient, and more aligned across every level of the organization.

The Increasing Importance of Digital Twins

The use of digital twins in manufacturing is steadily increasing in importance, bringing the physical and digital worlds together in ways that were once unimaginable. What started as simple 3D models for design and testing has grown into something far more powerful. Today, digital twins serve as dynamic, real-time reflections of everything from individual machines to entire supply chains.

At its core, a digital twin is a virtual representation of a physical object, process, or system. But this doesn't mean a one-time static model. A true digital twin continuously updates itself using real-time data, learning from its environment, and providing insights that drive better decisions. It's a living, evolving counterpart to the physical world, offering a way to monitor, simulate, and optimize operations with incredible precision.

The impact of digital twins spans across all of manufacturing. At the most granular level, machine twins track the health and performance of individual assets, predicting failures and optimizing maintenance schedules before problems occur. Process twins replicate entire production flows, allowing teams to fine-tune efficiency, reduce waste, and test improvements virtually before applying them in the real-world. Factory twins go even further, capturing data from across an entire plant to optimize resource allocation, energy consumption, and workforce planning. And beyond the factory walls, supply chain twins provide a complete, real-time view of logistics, inventory levels, and market conditions, helping businesses anticipate and prepare for disruptions before they happen.

The next generation of digital twins will be differentiated by their ability to connect and coordinate across layers. A machine twin might detect a potential issue with a robotic arm, sending data to a process twin, which then adjusts the production schedule to minimize disruption. A factory twin might analyze these changes in the context of overall plant efficiency, while a supply chain twin ensures that material flow is adjusted accordingly. The result is a fully integrated, intelligent system that reacts to challenges and opportunities in real-time.

These capabilities are becoming more critical as manufacturing grows more complex. Products are increasingly customized, supply chains are more volatile, and efficiency demands are higher than ever. Digital twins provide the intelligence needed to navigate these challenges. They allow manufacturers to simulate new production strategies before implementation, forecast equipment failures with high accuracy, and adapt operations dynamically based on demand and resource availability.

As artificial intelligence continues to advance, these virtual models will do more than reflect reality. Digital twins powered by AI will test countless production scenarios in parallel, identifying the best possible approaches before any real-world action is taken. They will work alongside AI agents to make automated decisions that go beyond what humans could achieve on their own.

To truly appreciate the intricacy that modern digital twins must capture, consider the complexity shown in this production flow diagram of a typical manufacturing environment:

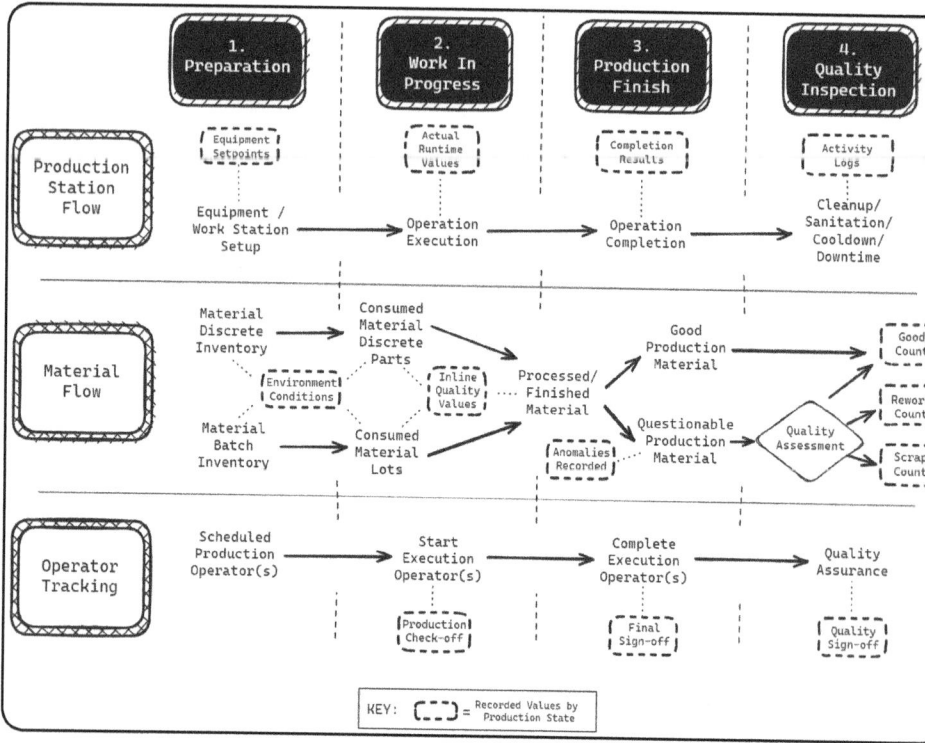

Figure 8-1 Process Flow Data Collection through a Production Operation by Station/Equipment, Material Transformation, and Operator

This diagram illustrates the complexity that modern digital twins must capture in a manufacturing environment. Every production station, material flow, and operator action generates multiple data streams that interconnect in ways that would be impossible for humans to track manually. The three parallel flows show how a single operation creates many decision points, quality checks, and potential variations. When digital twins connect all of these elements through AI analysis, they turn this overwhelming complexity into actionable intelligence.

The sheer volume of interconnected data streams reveals why artificial intelligence is essential for digital twins to reach their potential. AI systems can process these massive information flows immediately, detecting subtle patterns across all three flows simultaneously. If an operator adjusts machine settings, while a simultaneous environmental temperature shift affects material properties, AI can predict the combined impact and suggest corrections before quality issues emerge. This integrated analysis turns digital twins from monitoring tools into predictive collaborators.

Looking ahead, digital twins will evolve far beyond tracking current operations to actively reshaping how manufacturing works. AI-powered twins will continuously experiment with virtual scenarios, testing thousands of optimization strategies in parallel. They'll discover process improvements that people would never consider, like noticing that momentary micro-pauses on a conveyor belt two stations upstream improve a product's surface finish, or identifying optimal material flow patterns that violate traditional efficiency rules but deliver superior results.

The biggest breakthroughs will come from digital twins that connect everything including suppliers, production lines, logistics, and field operations. These networks turn insights from one location into coordinated actions across the entire system. They'll aid in speeding up innovation by testing thousands of scenarios when demand shifts or new regulations emerge. This kind of coordination drives smarter, faster, and more agile performance across the full value chain.

This shift will also change how products are designed and improved. Real-time data from the field will feed directly into the design process, triggering updates that flow into production without slowing anything down. If a digital twin spots a way to boost performance, it can simulate the fix, adjust the process, and roll out the change across the network. The result is a loop where every product gets smarter with each version.

The fusion of digital twins, AI, and autonomous systems represents manufacturing's evolution toward true intelligence. This leads to fundamentally smarter operations that learn, adapt, and improve faster than systems relying on manual human insight ever could. As digital and physical worlds merge completely, manufacturers will gain capabilities that seem almost impossible today, turning uncertainty into competitive advantage through the power of intelligent prediction and response.

The Future of Augmented Reality in Manufacturing

As technology advances, information is no longer locked inside dashboards or buried in reports. It's moving into the workspace, where it can be used right away. Modern production systems generate a constant flow of real-time data, but that data only becomes valuable when people can access it quickly and act on it without disruption. Augmented reality makes this possible by placing live insights directly into the field of view of those on the factory floor. Operators no longer need to step away to check a screen or sort through paperwork. Instead, they can see exactly what matters, directly on the machine, tool, or part they're working with.

By layering information over the physical environment, augmented reality changes how people interact with equipment and processes. It removes the guesswork from routine tasks, helps prevent mistakes, and reduces reliance on people's memory or past experience. Maintenance crews can now identify problems faster, operators get visual instructions as they work, and engineers can walk the line and see system health at a glance.

This shift goes beyond convenience. It reshapes how knowledge is delivered and how fast new skills can be developed. New team members can follow guided steps as they work, reducing training time and improving consistency. Experts in one location can support repairs in another, sharing a view through AR glasses and offering real-time guidance. AI assistants running in the background highlight risks, suggest adjustments, and keep teams ahead of potential failures. What used to take hours of review and analysis now happens in seconds, right where the work is being done.

Consider a maintenance worker approaching a machine. Instead of consulting reports or waiting for diagnostic results, they see performance metrics floating above the equipment, along with any AI-generated recommendations for preventive action. If a failure is likely, they access a detailed breakdown of the issue with step-by-step repair instructions. If human expertise is needed, an off-site specialist connects instantly, providing guidance as if they were physically present.

Quality control stands to be revolutionized by this technology. Instead of relying solely on human judgment, augmented systems overlay tolerance ranges against the recorded values directly onto parts, helping inspectors instantly detect deviations. The ability to visualize defects at a microscopic level or compare real-time production output against digital blueprints drives higher accuracy and reduces waste.

For assembly operations, digital overlays provide real-time guidance tailored to the exact product being built. As workers move through the process, step-by-step instructions appear in their field of view, highlighting components, displaying torque specifications, and verifying each action before moving to the next. This interactive workflow reduces assembly errors, accelerates training for new products, and ensures consistency across production lines.

One of the most immediate applications is in inventory management, where smart visualization tools guide workers through picking and storage tasks with greater speed and accuracy. Operators scanning a warehouse see stock levels overlaid onto shelves, receive optimized routes for retrieving materials, and get real-time updates on shipments.

As AI systems grow more advanced, their ability to analyze patterns and make recommendations will have an even greater impact. Computer vision will automatically recognize components, detect wear, and flag quality issues, overlaying recommendations in real-time. Voice commands and natural language processing will make accessing information as easy as asking a question. Future applications will include fully interactive workspaces where AI-driven insights, human expertise, and automated systems work together in a dynamic, real-time environment.

Collaboration across locations will reach new levels of effectiveness. Remote specialists will see exactly what an on-site worker sees, overlaying guidance, annotations, and repair steps onto their shared view. This eliminates the need for extensive written instructions or troubleshooting calls, allowing experts to assist multiple sites without travel delays.

This shift changes how sites function by removing the divide between digital systems and physical workspaces. Real-time data, AI insights, and augmented visuals come together to create a smarter, more interactive environment. By placing real-time data directly in front of operators, manufacturing becomes more agile and efficient.

To make these innovations a reality, companies must ensure their data infrastructure supports real-time integration. Augmented visualization is only as effective as the data driving it. Systems need seamless connectivity, ensuring that AI-powered insights, sensor readings, and operational data flow instantly to workers who rely on them. Edge computing will play a big role in supporting this, minimizing delays by processing data closer to where it's generated.

Future advancements will push this integration even further. AI-powered AR systems will learn from experience, customizing information displays to fit individual workflows. They will predict operator needs based on patterns, highlighting relevant data before it's even requested. The result is augmented intelligence rather than simply augmented reality, creating seamless collaboration between the worker's skill and AI guidance.

Robotics Evolve to Physical AI Agents

Industrial robots have long been reliable tools known for being precise, efficient, and predictable. But they have been bound by rigid programming, unable to adjust beyond their predefined instructions. However, that reality will soon be changing. The next generation of robotics is emerging, bringing a shift from automated machinery to intelligent, adaptive agents that can reason, learn, and collaborate in real-time.

Think of this as Agentic AI in physical form. Just as AI-driven digital assistants make decisions based on data access and logical constraints, intelligent robots must operate within physical, spatial, and control boundaries. They can only act within the areas they have access to, use the data they are permitted to process, and make decisions aligned with their assigned tasks. This structured autonomy allows them to work alongside humans, adapt to changing conditions, and interact with other machines in ways that were previously impossible.

Today's manufacturing robots are built for repetition. A welding arm on an assembly line can deliver thousands of perfect welds in a shift, but even a small change like a misaligned part or unexpected material can throw it off. When that happens, the system either keeps going incorrectly or stops and waits for help. Collaborative robots, or cobots, offer more flexibility and can work safely alongside people, but they still operate within fixed rules. They respond to changes but don't adapt on their own. The same holds true for mobile robots in factories and warehouses. They can follow set paths and avoid obstacles, yet they don't adjust plans based on real-time demands across the floor.

Emerging robotics technology breaks new ground. These intelligent machines move beyond programmed instructions to incorporate synthetic thinking into every task. Using advanced AI models, real-time sensor data, and localized computations, they analyze their surroundings, adapt to unexpected conditions, and collaborate with other robots and human operators alike.

Consider a robotic arm assembling a complex product. Instead of blindly following a preprogrammed sequence, it recognizes when parts are slightly out of place and adjusts its grip accordingly. If a component is missing, it doesn't just stop doing its work. It queries an AI system for the best alternative action. This could mean retrieving a new part, adjusting the assembly sequence, or notifying an operations manager.

This shift creates a system where machines operate as autonomous but controlled agents, each making decisions within their own physical and logical boundaries. These robots won't replace human workers but will act as intelligent teammates, expanding the capabilities of the modern manufacturing floor.

At the individual level, AI-driven robots can make real-time adjustments, solving problems as they arise. Imagine two robotic arms assembling a product together. In a traditional system, their movements would be preprogrammed, requiring strict coordination to avoid collisions. If one robot encountered a delay, the other would have no way to react dynamically. Intelligent robots, however, can sense each other's actions and adapt. If one slows down due to a temporary issue, the other can adjust its sequence accordingly. If a high-priority task emerges, the robots can apply the rules they've been given to decide among themselves which one is best positioned to take it on without waiting for human input.

The same level of autonomy applies to mobile robots managing material flow in warehouses or on the production floor. Rather than following fixed routes, these machines can continuously optimize their paths and task assignments based on real-time demand. If a sudden spike in orders requires certain materials to be delivered faster, the robots can reorganize on their own, shifting priorities without needing direct commands from a central system. This kind of self-directed coordination is what moves traditional automation into something far more intelligent and adaptable.

Beyond working with other machines, these systems are also changing how robots interact with people. Traditional collaborative robots rely on simple safety mechanisms, like stopping when a human gets too close. Future robots will go further. Using AI to analyze human movement and predict intent, they can anticipate when an operator is about to reach for a tool or move into a workspace. Instead of merely reacting, they will proactively adjust their position, making the workflow smoother and more natural.

Safety is a critical factor in this evolution, but AI is shifting the approach from rigid physical barriers to continuous, intelligent risk assessment. These robots won't just stop when someone enters their area. They will analyze the situation, recognize potential hazards, and modify their actions in real-time to keep the environment safe. If a robot detects that one of its components is wearing down and could lead to an unsafe operation, it won't wait for a failure. It will slow its movements, adjust its precision, and schedule its own maintenance before the issue escalates.

This adaptability extends beyond safety. AI-driven robots won't need to be reprogrammed every time a new product variant is introduced. Instead of following fixed routines, they'll understand the principles of their tasks. If a part changes slightly or a new material is introduced, they'll adjust their strength, welding technique, tooling, or assembly sequence automatically. This flexibility is critical for manufacturers shifting toward more customized production, where shorter runs and rapid product changes are becoming more common.

These individual capabilities become revolutionary when robots work together as a coordinated team powered by centralized AI coordination. When robotic systems are connected to broader manufacturing intelligence, they can anticipate shifts in demand and reconfigure themselves accordingly. If the system predicts a surge in orders for a specific product, robotic work cells can start adjusting production parameters before the increase even happens. Material handling robots can reorganize their workflows to ensure the right components are available at the right stations, reducing bottlenecks before they occur.

Consider a scenario where a production line encounters a defect issue that requires adjusting both assembly sequences and material movement. A traditional robotic system would require human intervention to pause production, reprogram machines, and manually restart operations. In an AI-driven environment, the robots would detect the issue, propose alternative assembly sequences, and coordinate material handling adjustments automatically. They would validate these decisions against safety and quality parameters before executing them, ensuring minimal disruption while maintaining high standards.

The rise of autonomous robotics brings new challenges in oversight and governance. As machines take on greater responsibility, manufacturers must ensure that efficiency doesn't come at the cost of safety, security, or ethical integrity. How much independence should robots have? When should they seek human approval before making changes? These questions aren't just theoretical, they're central to shaping the future of how AI-driven robots will be incorporated in manufacturing. This means designing layered validation systems where robotic decisions are continuously checked against real-world constraints before being executed.

Training and certification play a big role in progressing this technology. Just as human workers must be trained and certified for specialized tasks, AI-driven robots need rigorous validation before they operate autonomously. Manufacturers must test and refine AI models to ensure they can handle both routine tasks and unexpected edge cases.

These intelligent machines will transform factories into adaptive ecosystems where robots and humans collaborate as partners, not just coworkers. Machines will learn from experience, share insights across networks, and adapt to new challenges without direct programming. The most advanced manufacturing environments will combine AI-driven robotics with human expertise, ensuring that efficiency, safety, and quality remain at the forefront.

With the right frameworks in place, AI-driven robotics will elevate production into a dynamic, adaptive process, pushing the boundaries of efficiency in the modern factory.

AI-Integrated Wearables and Human-Machine Collaboration

While augmented reality has already changed how workers engage with their surroundings, the next step is deeper integration between people, AI, and robotics. Wearable technology is driving this shift by providing real-time information and insights, enabling workers to interact effortlessly with the digital ecosystem of the factory. Intelligent platforms are incorporating workers' actions and feedback into a comprehensive, site-wide perspective, creating a more connected environment.

Imagine an environment where AI understands worker fatigue, anticipates safety risks, and adjusts workflows dynamically. Not for surveillance, but to ensure that every person is supported, protected, and working at peak performance. These technologies amplify human capabilities, allowing them to work safely and efficiently alongside increasingly autonomous systems.

The convergence of biometric sensors, edge computing, and AI is reshaping how people and machines work together. Wearable devices continually evolve to influence real-time decisions concerning task assignments, production flow, and robotic coordination. This creates a workplace where automation adapts to people, not the other way around. AI-driven wearables can track muscle strain and request robotic assistance before fatigue sets in. Smart gloves with sensory feedback can guide workers with precision beyond natural human ability. Exoskeletons adjust to an individual's movement patterns, reducing strain and optimizing ergonomics.

Consider a technician approaching a faulty machine. Their wearable has already pulled up maintenance history, analyzed sensor data, and highlighted possible causes. If the AI detects stress or fatigue in the worker's biometrics, it adjusts the pace of the procedure to prevent mistakes. Meanwhile, on the assembly line, exoskeletons reduce strain on workers lifting heavy components, while AI monitors movements, adapting support to each person's individual needs.

Biometric sensors take this even further. Clothing embedded with smart textiles can monitor heart rate, hydration levels, and cognitive alertness. When a worker shows early signs of exhaustion, AI can reassign tasks, adjust work schedules, or recommend short breaks before safety or efficiency are impacted.

Instead of simply detecting potential issues, these systems work proactively to prevent them. If AI senses rising stress levels in an operator performing precision assembly, it doesn't just raise an alert. It adjusts the production environment. A conveyor might slow slightly, or the spacing between incoming parts might increase just enough to reduce pressure without affecting overall production. These subtle, real-time adaptations keep workers performing at their best while maintaining operational flow.

This predictive capability extends to workforce management. AI can track patterns in biometric data, recognizing early signs of fatigue before they become visible. By analyzing this alongside production schedules, skill levels, and task demands, it can optimize shift transitions and task assignments to prevent exhaustion. Instead of waiting for productivity dips or safety incidents, the system continuously balances workloads to keep operations running smoothly.

Advanced touch-based feedback is taking human-AI collaboration even further. Smart gloves provide real-time tactile guidance that goes beyond simple vibration alerts. A technician wearing AI-enhanced gloves might feel a subtle shift in resistance, gently guiding their hands to apply the correct pressure when securing a component. Instead of looking away to check a manual, they can work with precision through intuitive physical cues. The system learns each worker's strengths, preferences, and fatigue levels, adjusting its guidance accordingly.

This kind of AI-driven physical assistance is reshaping how workers gain new skills. With wearable feedback systems, employees can adapt to new tasks or product variations quickly, without requiring extensive retraining. The AI provides in-the-moment coaching, reducing errors and allowing manufacturers to deploy workers more flexibly across dynamically changing roles.

As these technologies evolve, workers equipped with smart wearables become an important part of the factory's intelligence network. Every movement, action, and decision feeds into an adaptive system that continuously refines workflows, improves safety, and enhances collaboration between humans and machines. When AI can model human movement with the same precision as machine data, it can significantly optimize them. This awareness reshapes how people and robots collaborate. Wearables equipped with motion sensors, location tracking, and biometric feedback ensure that robots don't rely solely on their onboard cameras or proximity sensors to detect human presence. Instead, they receive a continuous data stream that allows them to react proactively.

The ability to predict and adapt in real-time extends beyond safety to how tasks are assigned and adjusted dynamically. If a technician is called away for an urgent repair, the system calculates the impact on surrounding processes and adapts. Robotic assistants may be redirected to help with tasks normally handled by the technician, or nearby workers might receive subtle adjustments to their workflow to compensate. Production speeds may shift slightly to maintain quality, ensuring that the workload remains balanced without overwhelming the team.

Powered work-assist devices like smart exoskeletons take this integration even further. These devices actively communicate with AI systems to anticipate and help with physical tasks. When a worker moves to lift a heavy component, their exoskeleton already knows the weight, center of gravity, and optimal lifting path of the item by combining real-time data with material specifications. It provides just the right amount of support while still allowing natural movement. In a fully integrated system robots also assist by holding parts in place for assembly, adjusting their positioning to reduce worker strain, or synchronizing their speed with the worker's pace.

In highly complex environments, AI-driven wearables enable predictive task orchestration. If a system detects that an operator will need a specific tool in 30 seconds based on the current workflow, a robotic assistant can automatically retrieve and position it in advance. If a worker is about to enter a hazardous zone, their smart vest triggers adaptive safety measures, like reducing the speed of nearby robots or activating visual guidance to highlight potential risks.

Real-world examples of this technology are already emerging. In some advanced facilities, workers wearing sensor-equipped gloves or headsets can control nearby robots with simple gestures. A technician might point to a tool, and an AI-powered assistant seamlessly retrieves and presents it. These systems don't just follow commands, they understand the context of the task. They recognize which tools are needed for different procedures, adjust their movements to match the worker's pace, and ensure smooth handoffs that minimize disruption.

This degree of integration opens up new opportunities for process improvement. AI can analyze data across hundreds of shifts, identifying subtle differences between top performing teams and those that struggle. The insights gained might reveal that something as simple as swapping two assembly steps or repositioning a supply cart can significantly improve efficiency and reduce strain. Instead of relying on intuition or trial and error, manufacturers can make evidence based changes that enhance both productivity and worker satisfaction.

By integrating data from human movement with AI automation, the factory floor evolves into a fully adaptive ecosystem where people and machines understand and respond to each other in real-time. Robots don't just react, they anticipate. Workers don't just follow instructions, they collaborate with intelligent systems that amplify their skills. The result is a workplace that's safer, more efficient, and more responsive than ever before.

The Impact of Quantum Computing

A new era of computing is emerging, and it is one that will dwarf anything we've seen before. Quantum computing is more than just an improvement over classical computing, it's a completely different way of solving problems. Where traditional computers process information as a sequence of ones and zeros, quantum computers operate in a more complex state, capable of handling multiple possibilities at once. This allows them to solve problems in minutes that would take even today's fastest supercomputers hundreds of years to complete.

The implications of this for manufacturing are far-reaching. Quantum computing has the potential to break through the limits of classical optimization, simulation, and pattern recognition. Problems that were previously too complex to solve suddenly become feasible. This includes complex problems like predicting every possible disruption in a global supply chain, optimizing production schedules in real-time while considering the impacts of each work cell, or simulating the molecular behavior of new materials to optimize how they can be incorporated in manufacturing processes.

Artificial intelligence is already reshaping manufacturing, but quantum computing will advance its capabilities even further. AI needs vast amounts of data, yet training sophisticated models can take weeks or months using conventional hardware. Quantum computers could process massive datasets far faster, revealing deeper insights and making real-time AI suggestions more powerful.

Quantum-enhanced AI will reimagine how products are designed for both performance and manufacturability. Today, engineers face trade-offs between durability, weight, cost, and ease of production. Optimizing for one factor can create challenges for another, requiring extensive testing and iteration. Quantum computing changes this equation by analyzing countless design possibilities simultaneously, factoring in both product performance and how efficiently it can be manufactured.

Instead of just finding the strongest or lightest design, quantum systems can evaluate how different shapes, materials, and assembly methods affect production speed, energy use, and material waste. A product that once required multiple complex steps to assemble might be redesigned in a way that simplifies automation, reduces defects, or minimizes costly materials. This level of design intelligence could revolutionize industries that rely on precision engineering, from automotive assembly to high-tech electronics.

Manufacturing is filled with optimization challenges such as scheduling thousands of machines, coordinating fleets of autonomous robots, or managing energy use across massive facilities. These are all areas where even slight improvements can yield enormous gains. Today's methods rely on approximations and educated guesses because the full range of possibilities is simply too vast to compute. Quantum algorithms will eliminate those limitations.

Consider optimizing a global supply chain that involves millions of variables such as shipping times, weather patterns, supplier delays, geopolitical risks, fuel prices, warehouse capacity, local regulations, and countless other factors. Classical computing struggles to account for these complexities in real-time. Quantum computing could instantly analyze every possible disruption and recommend the best alternatives before problems arise. The result would be supply chains that stay efficient while remaining resilient and ready to adapt to disruptions before they cause delays.

Even on the factory floor, quantum-based scheduling enables fully adaptive production lines. Instead of static, preplanned schedules, AI-driven systems could coordinate machine tasks based on real-time demand, material availability, and worker performance. Quantum computing can process these variables collectively, identifying optimal solutions in a fraction of the time required by traditional computing. Autonomous systems in manufacturing will also become more intelligent and responsive, with robots and AI-driven logistics systems using quantum-powered algorithms to make split-second decisions with greater accuracy.

The rise of digital twins has already impacted manufacturing by allowing companies to simulate and test processes before making real-world changes. But even the most advanced digital twins today are limited by computing power. Quantum computing will remove those limits. Instead of simulating production processes one variable at a time, quantum-enabled digital twins could analyze thousands of variables at once, uncovering inefficiencies and predicting outcomes with near-perfect accuracy.

In a high-precision manufacturing environment, minor variations in temperature, humidity, and vibration can affect product quality. Quantum-powered digital twins could process all these environmental factors in real-time, making micro-adjustments to prevent defects before they occur.

Quantum computing won't replace classical computing overnight. The first impact will come from hybrid systems, where quantum processors handle the most complex tasks while traditional systems manage everyday operations. AI and optimization algorithms will need to be rewritten to take advantage of quantum capabilities, and manufacturing systems will need new interfaces to integrate with quantum insights.

This new level of computational power will completely revolutionize the materials science space. Developing stronger, lighter, or more heat-resistant materials currently requires extensive trial and error. Simulating molecular behavior is so computationally intensive that even the most powerful computers must simplify calculations, limiting accuracy. Quantum computing changes this by processing molecular interactions in ways that directly mirror the physics of the real-world. This could accelerate the discovery of new materials for aerospace, automotive, and electronics manufacturing, enabling breakthroughs in performance, durability, and sustainability.

Machine learning models will also evolve as quantum computing enables AI to analyze complex, high-dimensional datasets in ways never before possible. Predictive maintenance, quality control, and demand forecasting will all improve as AI systems gain the ability to detect patterns across massive datasets in real-time. Instead of waiting for a machine to show signs of wear, AI enhanced by quantum processing could predict failures long before they happen by analyzing sensor data, historical trends, and external factors simultaneously.

The benefits of how quantum computing will impact manufacturing are limitless. The ability to solve optimization problems in seconds, create materials with unprecedented properties, and implement AI-driven automation will reshape the industry. Companies that understand this shift and begin adapting their data architectures today will be the ones driving the next wave of industrial transformation. The question is no longer if this technology will change manufacturing, but how quickly businesses can position themselves to take advantage of it.

Insights in Action

Norman adjusted the AR safety glasses perched on his nose, still getting used to the feel of them. When Rita had introduced the new augmented reality system last month, he'd been skeptical. "I've been running these machines for years," he'd told her. "I can hear when something's not right."

Rita just smiled. "Give it a try, Norman. You might be surprised."

Now, as Norman walked down the production line, he had to admit she was right. The AR overlay was like having x-ray vision and a quantum computer rolled into one. As he glanced at each machine, real-time data floated in his field of view, including production rates, temperature readings, vibration levels, and more.

Suddenly, a red outline caught his eye. He turned to look at the large mixing vat at the end of the line. The AR display showed a pulsing red icon next to the temperature readout.

"That's odd," Norman muttered. He walked over to the vat and placed his hand on its side. It felt normal. Without the AR, he might have just walked by.

Norman tapped the side of his glasses, bringing up more detailed information. The temperature inside the vat was fluctuating more than usual, but still within the acceptable range. That's why the regular alarms hadn't gone off.

He peered into the vat, watching the mixture churn. Something caught his eye. There was a slight irregularity in the flow. Norman tapped his glasses again, this time pulling up the maintenance history of the agitator.

The AR display showed that the agitator's motor was drawing more power than usual, and its efficiency had been slowly declining over the past week. The system had noticed the pattern and flagged it, even though it hadn't yet affected product quality.

Norman opened up the diagnostic app on his glasses and ran a quick check. The results confirmed his suspicion, the agitator bearings were starting to wear.

With a few more taps, Norman scheduled a maintenance job for the upcoming weekend shift. As he did, a notification popped up in his view: "Similar wear patterns detected in Mixing Vats 2 and 4. Schedule preemptive maintenance?"

Norman chuckled. "You're getting smarter every day," he said, approving the additional maintenance.

As he turned away, he saw Rita walking towards him, a tablet in hand. "Hey Norman, I just got an alert about some maintenance you scheduled. Everything okay?"

Norman grinned. "Everything's just fine, Rita. Your fancy AR system just helped me catch a problem before it became one. These glasses are like having robotic eyes!"

Rita beamed. "Robotic eyes? I like that. You caught three problems before they could cascade into real issues. That would have meant hours of downtime next month." She glanced at her tablet. "Actually, make that four problems. The system just flagged a similar pattern in the packaging line based on your diagnosis."

As they walked together down the production line, Norman marveled at how the familiar factory floor now pulsed with visible streams of data. It was like seeing the heartbeat of the entire operation. He realized that this blend of his hard-earned experience and cutting-edge technology was opening up possibilities he'd never imagined.

Chapter 9

Steps to Building an AI-Enabled Manufacturing Enterprise

||||||||||||||||||||||||||||||||||||||||||||||||||||||||||||||||||||||||||||||||||||

Introduction

Technologies that once seemed like science fiction are now becoming everyday tools. AI-powered robotics work alongside people, models simulate entire production lines, and predictive analytics prevent breakdowns before they happen. Recognizing these changes is only the beginning. The real challenge lies in turning innovation into lasting improvement. Manufacturers need to move beyond scattered pilot projects and build fully integrated, data-driven operations that keep pace with emerging trends.

That's what this final chapter will cover. Throughout this book, we've explored the technologies, strategies, and architectures that are required to support modern manufacturing. We've addressed the challenge of system fragmentation and shown how AI, automation, and predictive insights can increase performance. The Unified Manufacturing Data Architecture provides the foundation to bring it all together, creating a single, trusted source of information across an organization. Now, it's time to shift from vision to execution.

Implementing an AI-driven manufacturing environment is a process that requires careful planning, cross-functional collaboration, and a structured approach. This chapter provides that roadmap. We'll walk through the essential steps of UMDA deployment, starting with assessing business needs and integrating data sources. From there, we'll focus on security, governance, and ensuring long-term scalability. At each stage, practical insights are highlighted to help manufacturers deploy technology in ways that create measurable impact along the journey.

Lasting change doesn't come from adopting the latest tools or chasing industry trends. It happens when data strategies align with real business needs. What are the biggest bottlenecks slowing production? Where are inefficiencies driving up costs? How can smarter insights improve quality, reduce downtime, or optimize resource use? Answering these questions will help guide the foundation of a truly impactful data and AI strategy.

To achieve this, companies must overcome the challenge of fragmentation across systems and teams. Without a unified approach, even the best technology struggles to deliver.

Once these fundamental elements are in place, the benefits become clear. AI and advanced analytics move beyond theory and start delivering measurable change. Predictive maintenance reduces unexpected breakdowns. Digital twins optimize production before physical adjustments happen. AI systems dynamically adjust processes as conditions change. Success comes from delivering insights where they matter the most and can be acted upon.

For some companies, this means untangling legacy systems and years of fragmented data. For others, it focuses on taking an already connected operation and enabling scalable AI solutions through cross-domain frameworks. No matter where you're starting, the goal is the same. Build a foundation where information flows between machines, systems, and people in a unified manner.

The approach in this chapter adapts to different manufacturing environments while maintaining consistent governance and standards. Every organization faces unique challenges, but core principles stay constant. Identify the business needs that matter most, integrate the right data sources, ensure security and accessibility, help teams trust and use the data, and complete each step with scalability in mind.

This is the point where ideas become reality. The steps that follow will help turn planning into action, setting the stage for manufacturing operations that don't just keep up with change but help shape it.

Business Needs Assessment and Use Case Development

Every effective transformation begins with a clear understanding of the organization's most pressing needs. The chosen technology must directly support the specific challenges and opportunities faced in daily operations. Without well-defined goals, even the most advanced tools are unlikely to produce meaningful results.

To ensure the data strategy aligns with what matters most, organizations should gather insights from those who understand operations best. This involves interviewing stakeholders across teams including production supervisors, operators, maintenance staff, quality inspectors, supply chain managers, and executive leadership.

During discovery, teams should identify pain points that appear across multiple areas. If several departments struggle with downtime or quality issues, these problems represent significant opportunities for improvement. Success comes from avoiding vague goals and defining specific, measurable use cases. Rather than generic goals like "reduce costs," organizations should target clear objectives like "reduce unplanned downtime by 20% using predictive maintenance." Specific goals make it easier to measure success and maintain focus throughout implementation.

One of the biggest mistakes companies make is leading with technology instead of business needs. Organizations that invest in advanced data platforms or AI tools without a clear purpose, find they don't solve the right problems.

Another common misstep is excluding end users from the planning process. If operators, technicians, and frontline workers don't see value in the system or find it difficult to use, adoption will suffer regardless of technical capabilities.

A lack of clear success metrics can also derail progress. Without measurable outcomes, it becomes difficult to track improvements or secure ongoing support. Every use case should have quantifiable results, whether reducing downtime, improving yield, or cutting waste. This measurement framework becomes critical for demonstrating value and building momentum for broader initiatives.

After gathering insights, the next step is prioritizing them using three key criteria: business impact, feasibility, and strategic alignment. Business impact considers which use cases offer the biggest return on investment. Feasibility evaluates whether it's realistic to collect and apply the necessary data given current systems and capabilities. Strategic alignment ensures these initiatives directly support broader organizational goals and long-term vision.

This prioritized list becomes the foundation for the roadmap. It informs future technology and data decisions, ensuring that each step contributes to clearly defined outcomes. It also helps maintain alignment across teams by establishing a shared understanding of what success looks like and why specific initiatives take priority over others.

This initial assessment phase sets the entire project up for success. Taking time to clearly align with business needs builds confidence and momentum, making every subsequent step more effective and ensuring that the AI-enabled data architecture delivers measurable value from the start.

Steps to Success

Successfully implementing an AI-ready data architecture begins with identifying key business challenges, aligning stakeholders, and defining clear objectives to ensure the architecture delivers measurable value. The following steps outline how to achieve this:

- **Identify Critical Pain Points** – Analyze manufacturing inefficiencies, bottlenecks, and risk areas where better data utilization could drive improvement.

- **Engage Cross-Functional Stakeholders** – Bring together IT, OT, data teams, and business leaders to ensure alignment between technical capabilities and business goals.

- **Define Clear Objectives** – Establish specific, measurable goals such as improving equipment uptime by 28% or reducing defect rates by 12%, ensuring each objective directly addresses identified pain points.

- **Prioritize High Impact Use Cases** – Focus initial efforts on achievable projects that demonstrate immediate value by using the business impact, feasibility, and strategic alignment criteria to guide selection.

- **Establish Success Metrics** – Define quantifiable criteria such as efficiency gains, cost reductions, or increased throughput to track the impact of implementations and build momentum for broader adoption.

Infrastructure and Data Storage

Once business objectives are defined and use cases prioritized, the focus shifts to designing the infrastructure and identifying data storage requirements. It is essential to ensure appropriate storage solutions are in place for different types of data, along with efficient methods for moving information to where it is needed. The system must also offer the flexibility to scale alongside operational growth.

Before making technical choices, organizations should assess the current data landscape. How much data does the site generate daily? What types of data are involved including items such as simple sensor readings, detailed quality inspection images, or complex machine logs? How often is this data produced and how quickly does this information need to be accessed? A high speed production line might need real-time access to quality control data, while historical maintenance records can be stored in a lower-cost system that doesn't require instant retrieval.

One of the critical decisions in this process is choosing between on-premise, cloud, or hybrid storage. Think of this like deciding between owning a warehouse, renting space in a shared facility, or using a combination of both.

On-premise storage offers maximum control and performance for frequent, large-scale data transactions, but requires ongoing infrastructure investment and maintenance. Cloud storage provides scalability and flexibility with lower upfront costs, but security and latency may be concerns for time-sensitive applications. Hybrid storage combines both approaches, keeping sensitive or real-time data on local servers while using the cloud for less critical storage and analytics.

Many manufacturers find that a hybrid approach offers the ideal balance. For instance, a company might store real-time process control data on local servers to support immediate decision making while transferring historical performance data to the cloud for in-depth analysis.

A thoughtful data storage strategy will include both where the data is kept combined with how it's organized. A tiered storage approach ensures that information is placed in the right location based on how often it's accessed and how quickly it needs to be retrieved. Hot storage handles real-time data like current production metrics that need instant access. Warm storage manages recently used data needed for weekly or monthly analysis. Cold storage archives older data that must be retained for compliance or historical reference but is rarely accessed.

Edge computing plays a crucial role in this infrastructure by processing time-sensitive data locally rather than sending everything to centralized systems. This reduces latency for critical decisions while optimizing bandwidth and cloud costs. Edge processing works hand-in-hand with the storage strategy, keeping only essential data local while routing other information to appropriate storage tiers.

The Unified Data Layer serves as the transportation network that makes this distributed system work efficiently. It ensures that information flows seamlessly between different storage locations and processing points while maintaining context.

To maintain structure and governance across storage layers, data contracts define clear rules for how data is handled. These contracts ensure that as data moves between storage tiers and processing locations, it remains consistent, accessible, and properly formatted. They specify retention policies, security access, and quality standards, making it easier to manage data across the entire organization.

A common mistake in infrastructure planning is treating all data the same way. Not every piece of information needs to be instantly accessible or stored in the highest cost systems. Categorizing data based on frequency, urgency, and business value helps optimize storage costs while ensuring that performance is prioritized where it matters most.

Another consideration is scalability and cost management. Data volumes in manufacturing environments will grow faster than expected. A well-designed infrastructure anticipates this growth using modular approaches that allow for expansion without disrupting ongoing operations. Organizations should plan for both technical scalability and budget scalability, ensuring that growing data volumes don't create unsustainable costs.

Steps to Success

A scalable and secure infrastructure is essential for efficient data storage, retrieval, and processing while balancing cost and performance. The steps to achieve this are as follows:

- **Assess Current Data Landscape** – Analyze data volumes, types, access patterns, and performance requirements to understand infrastructure needs and growth projections.

- **Determine Storage Architecture** – Select the optimal combination of on-premise, cloud, and hybrid storage based on latency, security, scalability, and cost requirements.

- **Implement Tiered Storage Strategy** – Deploy hot, warm, and cold storage layers to balance accessibility, performance, and cost based on data usage patterns.

- **Integrate Edge Computing Capabilities** – Process time-sensitive data locally to reduce latency, minimize bandwidth usage, and optimize cloud costs for critical operations.

- **Deploy Unified Data Layer (UDL)** – Implement the data transportation network to ensure seamless information flow across storage tiers and processing locations while maintaining context.

- **Establish Data Contracts for Governance** – Define structured agreements for data access, quality, retention, and standardization across all storage and processing environments.

Data Source Identification and Integration

The next step is identifying and integrating the right data sources. At this point information from multiple sources are brought together into a unified framework by harmonizing scattered, stand-alone systems into a single data environment. Getting this step right ensures that advanced analytics and AI-driven solutions are built on accurate, complete, and accessible data.

Start by mapping out all available high-level data sources. Manufacturing environments generate information from a wide range of systems, including production tracking software, machine controls, quality assurance databases, and supply chain platforms.

Some of the most valuable sources include:

Enterprise Resource Planning (ERP) systems manage inventory, procurement, and broader business operations.

Manufacturing Execution Systems (MES) coordinate production workflows, equipment, and process steps.

Quality Management Systems (QMS) store quality results, inspection records, and compliance documentation.

SCADA systems monitor real-time process controls and machine performance.

Programmable Logic Controllers (PLCs) capture signals from equipment to track status and automation sequences, recording this data in Data Historians.

IoT sensors provide real-time machine health, material tracking, environmental conditions, and utility consumption data.

For each use case, determine which data sources are essential, considering both direct and supporting data. A predictive maintenance initiative, for example, will primarily use machine sensor readings, but integrating maintenance logs and production schedules could provide deeper insights into failure patterns. By thinking holistically, companies can extract more value from their data without adding unnecessary complexity.

Trying to integrate too many data sources at once can quickly become overwhelming. A more effective approach is to start with a well-defined set of high impact sources and expand gradually as the system matures. For example, integrating batch processing data, sensor inputs, and quality control logs can provide immediate improvements in production consistency. Once that foundation is in place and delivering results, additional sources like supply chain and environmental data can be incorporated to further optimize efficiency and reduce waste.

Data federation plays a crucial role in this integration strategy by enabling access to information across multiple sources without requiring physical data movement. This approach allows organizations to connect disparate systems while maintaining data in their original locations, reducing storage costs and complexity while preserving existing system performance. Federation becomes particularly valuable when dealing with legacy systems that cannot be easily modified or when regulatory requirements mandate that certain data remain in specific locations.

A major challenge in integration is aligning different systems that may use inconsistent terminology, data structures, or reporting formats. Establishing a common data ontology, where terms, metrics, and formats are standardized, prevents confusion and ensures that combined datasets make sense. When one system tracks "machine downtime" and another logs "equipment unavailability," these differences need to be reconciled so that reporting and analytics remain accurate.

Technical complexity is another hurdle. Legacy systems may use proprietary formats, inconsistent naming conventions, or lack modern integration capabilities. Bridging these gaps requires careful planning, flexible integration strategies, and tools that can standardize and translate data without disrupting operations.

Beyond technical hurdles, integration requires cooperation across departments that have historically worked independently. Clear communication about data ownership, access requirements, and integration timelines ensures that teams understand their roles in the process and can plan accordingly.

A strong data integration strategy lays the foundation for everything that follows. By taking a measured, systematic approach, organizations can build a unified data environment that supports both immediate goals and long-term success.

Steps to Success

Integrating diverse data sources into a standard architecture requires careful selection, standardization, and validation to ensure seamless interoperability and reliable data flow. The steps to achieve this include:

- **Catalog Existing Data Sources** – Identify and document MES, IoT sensors, SCADA, ERP, and other manufacturing systems that will feed or consume data, including their formats and access methods.

- **Prioritize High Impact Data Sources** – Select critical data sources that directly support priority use cases, avoiding complexity by starting with a manageable set before scaling.

- **Implement Phased Integration Approach** – Begin with core operational data sources and expand gradually as the system matures and delivers proven results.

- **Deploy Data Federation Architecture** – Enable access to distributed data sources without physical movement, maintaining data in original locations while providing unified access.

- **Create Common Data Model (CDM)** – Standardize how data is structured across different systems to ensure consistency and interoperability throughout the integration process.

- **Develop Flexible Data Pipelines** – Build adaptable data pipelines that can handle structured and inconsistent data from diverse sources while accommodating future expansion.

- **Implement Data Validation Rules** – Establish automated checks to maintain data accuracy, completeness, and consistency across all integrated sources.

Data Ingestion and Standardization

Data ingestion and standardization ensure that raw information is transformed into a structured, reliable flow that supports AI and advanced analytics. A clear approach helps maintain consistency in formats, naming conventions, and data quality, leading to accurate insights and smoother technology adoption across the Unified Manufacturing Data Architecture.

Creating automated pipelines forms the foundation of any Unified Data Layer. These pipelines pull together live streaming data, transactional events, and batch master data updates from across the enterprise. They handle everything from structured machine readings and system logs to unstructured operator notes and inspection photos.

The timing of data flow depends entirely on business needs. Quality sensors demand instant processing to catch defects the moment they appear, while long-term trend analysis works perfectly well with batch updates that run overnight when computing resources are plentiful.

Before any data enters the data platform, it should pass through validation checkpoints to catch errors early. Equipment logs may contain inconsistencies, sensors will drift, and manual entries are prone to typos. Without these safeguards, bad data can create misleading insights, affecting production efficiency and quality control. Automated validation routines flag anomalies, helping prevent costly mistakes before they propagate through the system.

To ensure interoperability, all incoming data should adhere to the established Common Data Models with standardized terminology and formats. Without this standardization, different systems may label the same metric inconsistently, making it difficult to analyze trends across departments. The CDMs ensure that production data, supply chain information, and quality metrics each follow their domain-specific structures.

Adding metadata tags brings order to complex data by attaching clear labels to each data point. These tags record important details like when and where something happened, which machine or batch was involved, who was on shift, and what the environmental conditions were at the time. This makes it much easier to find and connect the right information when it's needed. It also speeds up pattern detection and strengthens predictive analytics by linking related pieces of data across the production process.

For real-time use cases, event driven architectures ensure that critical data is processed immediately rather than waiting for scheduled updates. When a quality sensor detects an issue, the system can trigger immediate alerts, automatically adjust process parameters, and notify relevant personnel. When equipment shows signs of potential failure, the system can initiate maintenance requests and adjust production schedules. This changes data from a passive record into an active tool to assist process control and operational efficiency.

The complexity of data ingestion increases when integrating legacy systems with modern platforms. Many organizations face challenges with proprietary data formats, outdated interfaces, or siloed data sources. Addressing these issues requires a phased integration approach, starting with high value data sources and gradually expanding as systems are updated and the architecture matures.

A well-structured ingestion process improves the accuracy of analytics while making AI-driven applications more effective. When data is clean and consistently formatted within the UMDA framework, operators trust the insights, cross-department collaboration improves, and decision making becomes more proactive. These early investments in standardization create a scalable foundation that supports continuous improvement as needs evolve.

Steps to Success

Ensuring that data flows smoothly in a consistent and structured manner is critical to maintaining quality and usability throughout the unified architecture. The following steps outline how to achieve this:

• **Set Up Automated Data Pipelines** – Establish real-time streaming and batch ingestion processes based on use case requirements, balancing speed with resource efficiency.

• **Validate Data at Ingestion Points** – Apply automated rules and checks to identify errors, duplicates, and missing information before data enters the UMDA platform.

• **Apply Common Data Models** – Implement the established CDMs to ensure data uniformity within each domain while enabling cross-domain integration through the UDL.

• **Enrich Data with Manufacturing Metadata** – Tag data points with timestamps, equipment IDs, batch numbers, shift information, and environmental conditions for improved traceability and context.

• **Configure Event Driven Processing** – Implement message queues and streaming technologies to detect critical changes and trigger immediate responses to safety concerns, quality issues, equipment failures, and process deviations.

• **Execute Phased Legacy Integration** – Gradually bridge older systems into the UMDA architecture using the prioritized approach from source identification, minimizing disruption while modernizing the data infrastructure.

AI-Assisted Data Mapping and Contextualization

One of the toughest challenges in building a unified approach to data goes beyond creating new architecture. It involves making sense of years of existing information scattered across different systems. As new technologies were added over time, data was stored in various formats, with different names, and sometimes with different meanings. Trying to clean and align all of this by hand takes time and increases costs. AI helps accelerate the process by reading system records, spotting patterns, matching similar data, and organizing it into the established Common Data Models.

Bringing AI into data mapping starts with understanding what's already in place. AI systems can scan databases, spreadsheets, and log files to classify information, even if documentation is outdated or missing. It can recognize that "equip_stat" in one system and "machine_condition" in another are tracking the same thing. These systems can also detect inconsistencies, flag missing values, and suggest corrections to improve data quality before integration begins.

AI also simplifies schema matching, the process of figuring out how different systems connect. It can recognize that a customer ID in one platform ties to order data in another and quality checks in a third. That connection helps create the unified view needed for cross-functional insights. What starts as scattered data across systems becomes a clear, connected structure through the Unified Data Layer.

Data quality issues commonly surface during mapping. Duplicate records, conflicting values, and inconsistent formatting can all create problems. AI can identify patterns in errors, such as slight variations in equipment ID numbers or temperature readings recorded in different units, and suggest standardization rules to resolve them. This ensures that the resulting integrated model aligns with the CDMs and remains meaningful for real-world operations.

Context is critical when mapping manufacturing data to the appropriate domain models. A pressure reading of 50 means something completely different for a hydraulic press than for an HVAC system. AI models designed for manufacturing applications can recognize these differences, ensuring that production data maps to the Production CDM while environmental data fits the appropriate domain structure. These models learn from validation feedback, improving their accuracy over time and reducing the need for manual oversight.

The validation workflow should establish clear roles for workforce expertise in the mapping process. AI can propose mappings and standardizations, but experienced personnel must validate these suggestions to ensure they align with actual meanings and domain requirements. Production engineers should review equipment mappings, quality specialists should validate measurement classifications, and process experts should confirm that historical data interpretations remain accurate within the new framework.

One common mistake is trying to map everything at once. Tackling too much simultaneously can lead to delays and errors. A better approach is to prioritize datasets based on their importance to initial use cases, mapping critical production and quality data first before expanding to other sources. This phased approach aligns with the overall implementation strategy while building confidence in the AI mapping capabilities.

Another challenge is ensuring that historical data is interpreted correctly within the current CDM structure. What was considered "normal operating temperature" 20 years ago might not be accurate by today's standards. AI models designed for manufacturing contexts can recognize these shifts in standards and practices, ensuring that legacy records remain meaningful when integrated into the current data architecture.

Validating AI-generated mappings is critical for accuracy and trust. Cross-checking against historical trends and established benchmarks can reveal potential errors. If a mapped dataset suddenly shows an unexpected efficiency jump compared to past performance, it could indicate misalignment in the mapping process. Regular validation reviews ensure that mapped data remains accurate and supports reliable analytics across the platform.

Documentation becomes especially important when AI handles complex mapping decisions. Recording both how the data was mapped and the reasoning behind key choices helps maintain consistency and supports future governance. AI-driven mapping tools assist by automatically generating this documentation. The result is a clear record for audits, troubleshooting, and ongoing data management.

The goal of AI-assisted data mapping is to create a unified system where all data is clean, connected, and properly contextualized within the data architecture. When done correctly, this process builds an important component of the foundation required to support advanced analytics and AI applications across the organization.

Steps to Success

AI-assisted data mapping streamlines the process of integrating legacy information into the UMDA, making it accurate, consistent, and ready for AI applications. The steps to achieve this include:

- **Deploy AI Pattern Recognition Tools** – Implement AI systems to analyze disparate data sources, detect similarities in naming conventions, and identify relationship patterns across legacy systems.

- **Establish Manufacturing-Specific AI Models** – Deploy or configure AI models trained on manufacturing data structures, terminology, and domain contexts to ensure accurate mapping to appropriate CDMs.

- **Implement Structured Validation Workflows** – Create processes where domain experts review AI-generated mappings, with production engineers validating equipment data and quality specialists confirming measurement classifications.

- **Prioritize Critical Data Sources** – Focus initial AI mapping efforts on high value datasets supporting priority use cases, expanding systematically to other data sources.

- **Validate Data Quality and Accuracy** – Cross-check AI mappings against historical benchmarks and established patterns to identify potential errors before finalizing integrations.

- **Document Mapping Decisions and Logic** – Maintain comprehensive records of AI mapping rationale and human validation decisions to support governance, auditing, and future data management efforts.

Data Governance and Validation

Reliable AI depends on having data and calculations people can trust. Operators need to know that machine readings reflect what's really happening. Quality teams rely on accurate test results to keep standards high. Managers need solid data to make decisions that improve performance. As AI plays a bigger role in manufacturing, keeping data accurate, consistent, and secure will matter as much as delivering quality products. Strong governance works across the entire architecture, protecting data integrity from the moment it's collected through every stage of analysis.

Data governance provides the organizational structure needed to keep data usable and trustworthy across all domains and systems. Without clear ownership and validation rules, information can become fragmented, leading to errors and inefficient data usage. A strong governance strategy ensures that data remains reliable, accessible, and protected throughout the integrated architecture while supporting compliance with manufacturing regulations and industry standards.

The first step is establishing domain-based data ownership that aligns with the CDM structure. Production data requires stewards who understand manufacturing processes and equipment performance. Quality data needs oversight from specialists familiar with testing protocols and compliance requirements. Supply chain data demands expertise in logistics and vendor management. Each domain steward ensures that information within their area is recorded correctly, updated as needed, and used appropriately within the broader data architecture.

Access control implementation must balance security with operational efficiency across the integrated data environment. Role-based permissions prevent unauthorized changes while ensuring teams have the information they need. A maintenance technician may require machine performance data but shouldn't modify process parameters. A quality engineer might analyze testing data but shouldn't change calibration records. Production planners need visibility into multiple domains but may have read-only access to certain sensitive information.

Validation at the point of data entry helps catch early errors, but ongoing governance plays an equally important role in maintaining quality over time. Data contracts support this by defining the structure and format of the data and setting expectations for data quality, retention, and proper use throughout its lifecycle. These agreements help ensure that information stays reliable as it moves through the system.

Advanced AI-driven anomaly detection enhances ongoing validation by identifying subtle inconsistencies that emerge over time. These systems learn baseline patterns across domains and flag unusual trends, such as gradual sensor drift, unexpected correlations between production and quality metrics, or deviations in cross-domain relationships managed by the Unified Data Layer. This continuous monitoring builds on the AI mapping capabilities established earlier by using pattern recognition to maintain data integrity.

Tracking data lineage becomes essential for both operational efficiency and regulatory compliance in manufacturing environments. Organizations must demonstrate where data originated, how it flowed through the CDMs and UDL, what modifications occurred, and how it influenced decisions. This capability supports FDA regulations in pharmaceutical manufacturing, ISO standards in automotive production, and internal audit requirements across all manufacturing sectors.

Regular governance audits ensure that policies remain effective as the data architecture evolves and expands. Just as manufacturing teams perform preventive maintenance on equipment, data governance requires periodic reviews. Are domain ownership roles still appropriate as responsibilities shift? Do data contracts accurately reflect current operational requirements? Are validation rules catching relevant errors without creating bottlenecks? Governance frameworks must adapt to evolving functional needs and expanding AI capabilities.

Real-time governance monitoring becomes critical as organizations rely increasingly on data for immediate decision making. Waiting for periodic reports to catch data issues introduces operational risk. AI-driven validation tools continuously assess data quality as it flows through the integrated system, identifying problems immediately and triggering automated responses or alerts. This proactive approach minimizes disruptions and prevents inaccurate data from influencing critical manufacturing decisions.

A clear governance structure built into the data framework lowers compliance risks and builds trust in AI insights. When teams can rely on data from every system and domain, they act with greater speed and confidence.

Steps to Success

Strong governance and validation frameworks safeguard data integrity, security, and compliance across the integrated architecture, ensuring the system remains a trusted source of insights. The following steps will help achieve this:

- **Establish Domain-Based Data Stewardship** – Assign ownership for Production, Quality, Supply Chain, and other CDM domains, ensuring specialized expertise oversees each data area.

- **Implement Role-Based Access Controls** – Configure permissions that balance security with operational needs, ensuring teams can access required data through the UDL while protecting sensitive information.

- **Deploy Data Contracts for Ongoing Governance** – Implement data contracts that define data quality standards, retention policies, and usage guidelines that govern data throughout its lifecycle.

- **Configure AI-Driven Continuous Monitoring** – Establish anomaly detection systems that identify data quality issues, unusual patterns, and cross-domain inconsistencies in real-time.

- **Implement Comprehensive Lineage Tracking** – Maintain detailed records of data flow, transformations, and usage across CDMs and UDL to support compliance and operational auditing.

- **Schedule Regular Governance Reviews** – Conduct periodic audits of stewardship assignments, data contract effectiveness, validation rules, and compliance with manufacturing regulations.

Data Security and Compliance

Securing data is fundamental to ensuring operations run safely and smoothly without disruption. Every system, sensor, and machine that collects or transmits data across the architecture is both an asset and a potential risk. Striking the right balance between security and accessibility is critical to keeping production running while safeguarding valuable manufacturing intelligence.

The challenge grows as factories add IoT devices and edge computing to their operations. Processing data closer to the machines creates new points that must be secured, without slowing the real-time performance that makes edge computing useful. In some cases, these connected systems have exposed sensitive operational data, creating risks that can stretch from edge devices to central analytics platforms.

Traditional IT security methods don't always apply in manufacturing environments where information technology (IT) and operational technology (OT) systems must work together. Production environments require continuous operation, making it risky to apply security updates or patches that might disrupt manufacturing processes. Many industrial control systems were not originally designed with cybersecurity in mind, making them vulnerable if exposed to broader networks. Security measures must be carefully implemented to protect data flows between edge devices and central systems without interfering with real-time operations.

A layered security approach provides the best protection for distributed manufacturing data. The foundation starts with encryption, ensuring data remains secure both at rest in various storage tiers and in transit across the data architecture. Zero-trust principles add critical protection by requiring continuous verification of every request for data access, treating every user, system, and device as potentially untrusted until proven otherwise. This approach becomes especially important when securing access to federated data sources and cross-domain analytics capabilities.

Data contracts play a crucial role in enforcing security policies by defining not only data structure and quality requirements, but also access permissions and security classifications. These contracts ensure that security requirements travel with the data as it moves between systems, maintaining protection across the entire data lifecycle from edge collection through central analytics.

Security strategies in manufacturing are shaped by the specific regulations that apply to each industry. Manufacturers must follow data privacy rules like GDPR, cybersecurity frameworks such as NIST for critical infrastructure, and standards like ISO 27001. Defense suppliers need to meet requirements under NIST 800-171, while pharmaceutical operations must align with FDA rules for data integrity. These guidelines set clear expectations for protecting sensitive information and keeping operations compliant across complex, distributed environments.

Security monitoring needs to be proactive and automated to keep up with the speed and complexity of modern demands. AI-powered tools track patterns across local equipment, connected data platforms, and enterprise systems to detect unusual behavior as it happens. If a production line tries to access information it normally wouldn't, or if local devices start communicating in unexpected ways, the system triggers alerts immediately. This allows teams to step in before small concerns become serious risks.

Edge security requires special attention as more processing moves closer to production equipment. Simple edge devices have limited security capabilities but handle sensitive operational data. Implementing secure communication protocols, device authentication, and local threat detection ensures that distributed processing doesn't create security vulnerabilities. Edge security must integrate seamlessly with central security monitoring to provide comprehensive protection across the entire data architecture.

Disaster recovery planning is more complex in a distributed environment, but it's critical for keeping operations running. Plans need to cover local systems, shared data platforms, and the connections managed through the unified architecture. Secure, redundant backups should be stored in multiple places, with clear procedures in place to restore systems quickly. The aim is to make sure that security events don't lead to long production delays or lost information.

When done correctly, security supports manufacturing efficiency instead of holding it back. It ensures that people and systems get timely access to the data they need while guarding against new risks. Built into the data architecture, security works in the background, allowing operations to grow and adapt while keeping sensitive information safe across every system and site.

Steps to Success

Protecting against cyber threats while maintaining operations requires security measures designed specifically for distributed manufacturing environments. The steps to achieve this include:

- **Implement Zero-Trust Architecture** – Deploy continuous verification systems for all users, devices, and systems accessing data across edge, federated, and central environments.

- **Secure Edge Computing Infrastructure** – Establish device authentication, encrypted communication, and local threat detection for distributed processing capabilities.

- **Deploy AI-Driven Security Monitoring** – Implement automated threat detection that analyzes patterns of data changes from the factory floor to the cloud in real-time.

- **Enforce Security Through Data Contracts** – Integrate access permissions and security classifications into data contracts to ensure protection travels with data across the architecture.

- **Establish Distributed Disaster Recovery** – Create tested recovery procedures that account for edge devices, federated data needs, and UDL interconnections to ensure business continuity.

- **Conduct Regular Security Assessments** – Perform comprehensive evaluations of IT/OT security, edge device vulnerabilities, and compliance with regulatory agencies.

Data Accessibility and Utilization

Collecting data from different factory systems is challenging, but making it easy to find and use is even harder. True improvements come from turning raw data into accessible, actionable insights. In manufacturing, data is only valuable if the right people can obtain it quickly enough to make informed decisions. Without a well structured system, critical insights can get lost in reports, trapped in silos, or remain out of reach until it's too late to act on them.

Effective data access begins with knowing who needs which information and at what moment. Operators need immediate visibility into how their equipment is performing right at the machine. Quality engineers look for patterns across batches, pulling from data that spans systems and sites. Executives focus on big-picture insights that track factory performance across business units. Designing role-based access ensures that each person can access the data they need, in the context they need to see it, without being overwhelmed by irrelevant information.

The Edge Intelligence Hub plays a crucial role in making data accessible at the point of need. Rather than forcing operators to wait for central systems to process requests, the EIH provides immediate access to localized insights and real-time analytics. This local intelligence capability ensures that critical operational decisions can be made instantly, while still maintaining connection to enterprise-wide data through the Unified Data Layer.

Self-service portals have significantly enhanced how teams interact with data across the integrated architecture. Instead of waiting for someone else to generate reports, employees can explore real-time insights on their own terms through intuitive interfaces that connect to multiple data sources. If an operator notices an anomaly in production, they can immediately dig deeper using both local EIH capabilities and broader enterprise data without waiting for an analyst to investigate.

Real-time monitoring and AI-driven alerts help turn data into insights people can act on. A quality engineer might get an automatic warning about a small anomaly that could lead to defects down the line. A maintenance technician could be notified about a machine likely to need attention soon, based on both local sensor readings and broader maintenance history. These timely insights shift the focus from reacting to problems to staying ahead of them..

One common mistake is making data access too restrictive. While security is important, overly complex access rules can slow operations and frustrate employees. Tiered access models help balance security and usability by providing appropriate data visibility across the architecture. Operators see detailed, real-time performance metrics for their stations through local systems. Analysts get access to broader datasets for pattern recognition across domains. Corporate leadership receives high-level reports that incorporate insights from multiple locations and data sources.

A federated data architecture makes it essential to focus on how people search for and discover information. Users need to locate the right data whether it sits on local equipment, edge systems, or central platforms. AI-powered search tools help by finding what's needed across these sources, recognizing context and linking related data from different domains.

Training employees on how to use the platform is important but the major focus should be on empowering users to leverage the data effectively. Employees need to understand how to access data, interpret insights from different sources, and apply integrated analytics capabilities to solve day-to-day operational challenges.

The true measure of success in data accessibility is how quickly and effectively the workforce can leverage distributed data resources to drive positive change. When information flows seamlessly between edge and enterprise systems, operators can adjust processes in real-time, engineers can identify root causes faster using comprehensive datasets, and managers can make informed decisions that optimize performance.

Steps to Success

Making data accessible and actionable for all stakeholders requires well-defined access controls, intuitive interfaces, and self-service capabilities that work across the distributed architecture. The steps to achieve this include:

- **Define Role Specific Access Levels** – Establish tiered permissions that provide appropriate data visibility across edge, federated, and central systems without exposing unnecessary information.

- **Deploy Job Specific Dashboards** – Create customized interfaces that present relevant insights to operators, engineers, and executives using data from appropriate sources and domains.

- **Leverage Edge Intelligence Hub for Local Access** – Utilize EIH capabilities to provide immediate access to localized insights and real-time analytics at the point of operational need.

- **Implement AI-Powered Search and Discovery** – Deploy intelligent search tools that can locate relevant data across distributed sources, understanding context and relationships between different domains.

- **Enable Self Service Analytics Capabilities** – Provide intuitive tools that allow users to explore data independently across edge and enterprise systems without requiring IT intervention.

- **Configure Smart Alerts and Notifications** – Establish AI-driven alert systems that proactively identify trends and issues by analyzing patterns across local and enterprise data sources.

Data Audit and Inventory

A data audit provides a clear view of the information flowing through the architecture. It identifies what data exists, where it originates, and how it's being used across the organization. Without this visibility, inefficiencies go unnoticed, compliance becomes more difficult, and AI models may be based on outdated or incomplete information.

With factories generating data from hundreds of sources across a distributed environment, a clear structure is essential for keeping that information accurate, useful, and secure. The process starts by mapping what data exists in each domain, assigning ownership, and understanding how it flows through the architecture. Each dataset should have a responsible steward who ensures its quality, whether the data resides at a local site, in central repositories, or in the cloud. Modern audit tools support this structure by automatically recording every interaction with the data. They track who accessed it, when changes were made, and how it moved between systems. This creates a reliable record that strengthens security and supports governance.

Managing the data lifecycle becomes more complex in distributed architectures but remains crucial for optimization. Not all data needs permanent storage, and without clear retention policies, systems become bloated with outdated or duplicate information across multiple storage tiers. Some data must be retained for regulatory purposes, while other information may only be valuable for short periods.

AI-driven audits can simplify this process significantly by operating across the distributed data sources. Automated tools can detect duplicate records spanning multiple systems, identify outdated or redundant data across domains, and flag missing or incomplete records that could impact cross-domain analytics. Instead of manually sorting through massive datasets distributed across the architecture, AI continuously monitors and optimizes data quality in the background.

A centralized data catalog is key to managing complexity across the entire system. It provides a single reference point for data from factory sources to enterprise storage. This makes it easier for teams to find and understand the information they need. The catalog should map data relationships across the various CDMs and show how information moves through the unified data architecture. This helps teams see dependencies clearly and supports stronger collaboration.

Legacy systems hold valuable historical data, but bringing that information into a modern architecture takes thoughtful planning. Older platforms may use outdated formats or structures that don't align with current models. To handle this, map the data carefully so its context is preserved, standardize formats so the data fits into the new unified structure, and migrate it in stages to avoid disrupting daily operations. This approach helps ensure that critical insights from the past can contribute to accurate decisions for today.

A data audit is an ongoing process that must keep pace with evolving business needs, compliance requirements, and expanding capabilities. Regular audits ensure that AI solutions across the organization operate on accurate, up-to-date information, improving their reliability and effectiveness. By maintaining a clear, structured view of distributed data sources, organizations can enhance model training, optimize AI outputs, and ensure that insights remain relevant and trustworthy.

Steps to Success

Maintaining an accurate inventory of distributed data sources, ownership, and usage helps optimize storage, prevent duplication, and ensure compliance. The following steps outline how to achieve this:

- **Create a Comprehensive Data Catalog** – Document all data sources across factories and centralized systems, including ownership, purpose, and relationships between CDM domains.

- **Implement Automated Data Tracking** – Use tools that track how data moves and is used across the entire system, from edge processing to central analytics.

- **Deploy AI-Driven Audit Capabilities** – Use automated tools to detect duplicates, identify outdated data, and flag quality issues across all systems and storage tiers.

- **Define Distributed Retention Policies** –Define clear guidelines for managing data at every stage, taking into account how the data is used and the specific storage requirements that support those uses.

- **Execute Legacy Data Integration** – Systematically integrate historical data from legacy systems into the modern architecture while maintaining context and ensuring compatibility with current CDMs.

- **Conduct Regular Architecture-Wide Audits** – Perform periodic reviews of data inventory, quality, and compliance across all systems to maintain accuracy and relevance.

Analytics and Data Processing

Industrial data flows continuously, creating opportunities for improvement with every new data point. When processed and analyzed effectively across the unified architecture, this information leads to smarter decisions that enhance efficiency, quality, and overall performance by turning data into insights that drive action.

Manufacturing analytics operates at three different speeds across the distributed architecture, each serving a critical purpose. Real-time analysis happens at the site through the Edge Intelligence Hub, right where data is generated. This is essential for immediate decisions, such as detecting when a part is out of spec or a machine is overheating. Next, streaming analytics processes data flows through the UDL, tracking patterns over minutes or hours across different domains and alerting teams before small issues become big ones. Finally, batch processing digs into historical data from across all CDM domains to uncover long-term trends, optimize workflows, and improve predictive models.

The Edge Intelligence Hub modernizes how sites handle real-time analytics by bringing processing power directly to the location. Instead of sending every sensor reading to central systems, the EIH analyzes data at the source and makes immediate decisions. For instance, a quality sensor monitoring product specifications no longer collects data just to be used in a report. The EIH immediately determines if something is drifting out of tolerance and triggers automatic adjustments. This instant response minimizes waste, improves consistency, and enhances process control without requiring round-trip communication to central systems.

While real-time edge analytics provide immediate value, the most powerful insights come from combining current conditions with historical patterns across multiple domains. When an AI system suggests adjusting a process setting, it draws on data from Production CDMs, Quality CDMs, and historical performance trends accessed through the UDL. This cross-domain analysis enables predictions that factor in equipment status, quality patterns, supply chain conditions, and maintenance schedules.

Cross-domain analytics within the Unified Data Layer opens the door to deeper insights by connecting information from different parts of the organization. Production efficiency trends can be examined alongside quality results and supply chain details to reveal root causes that cross functional boundaries. This complete view supports optimization strategies that improve the whole manufacturing system rather than focusing on isolated parts.

One common mistake is trying to analyze everything at once without considering the distributed nature of the architecture. A tiered approach ensures that critical edge analytics happen instantly through the EIH, streaming analytics process cross-domain patterns within minutes, and batch processing handles complex historical analysis during off-peak hours. This distribution optimizes both performance and resource utilization across the system.

To make analytics truly valuable, visualization tools need to match the distributed structure of the data and the specific roles of those using it. Modern dashboards bring together information from local systems and cloud platforms, showing each person what matters most for their work. Operators get real-time insights about their equipment, quality teams see trends across processes, and managers track performance across sites and departments.

The purpose of distributed analytics is to support better decisions, not to flood the organization with reports. Analytics should grow in line with the architecture as it evolves. The strongest path starts with site-level insights through the EIH, expands to cross-domain analysis with the UDL, and eventually builds toward enterprise-wide capabilities. This steady progression ensures that each step delivers practical value while creating a solid foundation for advanced AI models as the system matures.

Steps to Success

Transforming raw data into valuable insights requires advanced analytics pipelines, AI-driven models, and real-time processing. The steps to achieve this include:

- **Deploy Edge Analytics Through EIH** – Implement real-time analysis capabilities at production locations for immediate decision making and process control without off-site dependencies.

- **Establish Cross-Domain Analytics via UDL** – Enable analysis that correlates data across Production, Quality, Supply Chain, and other CDM domains to identify systemic patterns and relationships.

- **Implement Tiered Processing Architecture** – Configure real-time edge analytics, streaming cross-domain analysis, and batch historical processing to optimize performance and resource utilization.

- **Enable Advanced Analytics Capabilities** – Prepare the architecture to support machine learning and AI applications through proper data flows, quality controls, and processing infrastructure.

- **Create Role-Based Analytics Dashboards** – Design visualization interfaces that present relevant insights from local and enterprise analytics based on user roles and responsibilities.

- **Establish Analytics Foundation for AI** – Implement data processing and analysis capabilities that provide the foundation for AI.

AI Model Deployment and Integration

The AI deployment phase is where the effort to collect and standardize data through the UMDA starts delivering real results. With strong CDMs, UDL connections, and distributed processing in place, AI can help reduce manual work, cut waste, improve quality, and fine-tune processes. Success depends on careful integration that ensures AI works as a tool supporting human expertise across the entire manufacturing operation.

The first step is to focus on where AI can deliver the most value using the data foundation already built. The strongest applications address real challenges identified early and build on the existing data flows established in Common Data Models and connected to the Unified Data Layer. Predictive maintenance works well by using sensor data from production equipment. Quality can improve through insights drawn from inspection data. Other opportunities include fine-tuning processes with input from multiple domains or forecasting supply needs with integrated operational data. The goal is to provide insights that people can trust and use to make better decisions.

The right deployment approach depends on the data, how fast it needs to be analyzed, and how quickly action is required. The Edge Intelligence Hub is ideal for AI models that support real-time decisions. These models process information locally, make immediate adjustments, and help catch issues before they grow into bigger problems. Cloud platforms work better for more complex tasks, like finding patterns across multiple sites or running large-scale optimization using data from the Unified Data Layer. Many manufacturers use a mix of both, relying on the edge for quick decisions and the cloud for deeper, enterprise-wide analysis.

Agentic AI builds on the foundations already in place, enabling AI systems to act autonomously within set boundaries rather than only offering recommendations. These agents can fine-tune process parameters when quality metrics shift, schedule maintenance using predictive insights, or adjust production plans as demand changes in real-time. When deployed through the Edge Intelligence Hub, Agentic AI delivers immediate, local decisions while staying aligned with enterprise data through the Unified Data Layer.

AI deployment is only valuable if people trust and use the systems effectively. The best implementations integrate seamlessly into existing workflows, providing clear recommendations with explanations that operators can understand within their domain context. AI-driven suggestions should be easy to verify against real-world outcomes using the same data sources that feed the models. When users can confirm correct predictions and flag incorrect ones, this feedback improves the system over time through structured learning processes.

Building trust requires transparency in how AI systems make decisions, especially when they access data across multiple domains through the UDL. Operators need to understand what the AI recommends, why it reached that conclusion, and which data sources influenced the decision. This clarity is even more important when AI recommendations affect multiple operational areas or when local AI systems coordinate with other sites.

Staged deployment remains the best approach for integrating AI across the unified architecture. Rolling out models in controlled environments first allows teams to test performance within specific CDM domains before expanding to cross-domain applications. Initial deployment might focus on single domain AI applications, such as quality prediction models using only Quality CDM data. Once these prove effective, deployment can expand to more sophisticated models that leverage data across multiple domains through the UDL. This phased approach prevents costly mistakes and ensures AI delivers value at each expansion stage.

Measuring AI effectiveness means using metrics that reflect how it works across distributed systems and diverse data sources. Key measures include how well predictions match actual results, how much time is saved through automated insights in different areas, and the cost savings from preventive actions. Quality gains from AI-driven adjustments, adoption rates across sites, and the return on investment for each application also provide valuable insights.

Deploying AI at this stage means putting the Feedback Data Layer into action as part of the full system. The FDL captures both AI-generated insights and human validation, creating a record of recommendations, decisions, and outcomes across the operation. It draws on data from the UDL, tracking how AI suggestions perform in practice and how operators and engineers respond. This continuous feedback loop helps refine models, combining human expertise with AI learning to improve accuracy and relevance over time. The result is a growing knowledge base that strengthens future decisions and keeps the system aligned with real-world performance.

AI should enhance human capabilities rather than replace them, particularly in the context of distributed manufacturing operations. The best results come from strong partnerships between machine intelligence and human expertise, where AI processes vast amounts of data from across the architecture while experienced operators provide context and judgment for final decisions.

Long-term success with AI starts by focusing on high-impact use cases that build on existing data flows. Models improve as they learn from real-world outcomes captured through the FDL. From there, AI can gradually expand across the distributed architecture. The goal is to treat AI as a continuous improvement tool that grows with the business while supporting smarter decisions across the company.

Steps to Success

Effective AI deployment leverages the established data architecture to deliver measurable value while building user trust and adoption. The steps to progression include:

- **Identify High Value Use Cases Within the Data Architecture** – Select AI applications that leverage existing CDM data flows and address business needs identified during initial assessment phases.

- **Choose Optimal Deployment Locations** – Determine whether AI models should run through the EIH for real-time decisions, in cloud systems for cross-domain analysis, or in hybrid configurations that optimize both approaches.

- **Deploy Edge AI Through EIH** – Implement real-time AI capabilities at production locations that can make immediate decisions using local data while maintaining connection to enterprise systems.

- **Implement Cross-Domain AI through the UDL** – Deploy AI models that analyze data across multiple CDM domains to identify patterns and optimization opportunities that span operational areas.

- **Establish Staged Validation Process** – Test AI models in controlled environments within specific domains before expanding to cross-domain applications and full-scale deployment.

- **Deploy a Feedback Data Layer** – Implement structured feedback collection that combines AI-generated insights with human validation, tracks operational outcomes, and continuously refines AI performance across sites.

Data Lineage and Traceability

Building trust in AI-powered manufacturing starts with clear visibility into every step data takes along its journey. Teams need to trace where data originates, how it's altered, and what influenced each recommendation. This becomes essential when quality issues arise, auditors ask questions, or teams want to validate AI decisions. Data lineage provides that clarity, tracking information from its source through every system that touches it.

In traditional setups, tracing data means digging through logs, matching timestamps, and hoping connections stay intact. The UMDA changes that by making traceability part of the system's DNA.

Each data object, whether raw, calculated, or aggregated, must carry references to its origin. This includes the source system, timestamp, transformation logic, and its role within the broader namespace (like production, quality, or maintenance). These lineage markers must remain intact as data moves through ingestion, modeling, processing, and analysis.

The journey begins at the edge, where context is woven in immediately. A temperature measurement becomes meaningful because it's tied to the specific machine, product, and material under defined conditions. That context is preserved by the local Common Data Model so that when data moves to central systems, its meaning travels with it. This foundation makes everything downstream more reliable and traceable.

As data flows through the architecture, each step of its journey is recorded. The Unified Data Layer maintains these connections while harmonizing the data across domains. When production figures combine with quality and maintenance data, those relationships remain intact. Each alteration or analysis simply adds another layer to the lineage, never erasing its origins.

The Feedback Data Layer completes the picture by capturing decisions and their outcomes. When an AI model suggests a process change, the FDL logs the input data, model version, confidence levels, human review, and actual result. With this record, the system learns over time while preserving full auditability.

Modern tools make lineage capture easier. Metadata platforms automatically track where data comes from, orchestration tools record how it is processed, and model registries log version history. When built into the UMDA, lineage becomes a natural part of the process instead of a burden.

Data contracts help by specifying what metadata must accompany each exchange, such as data source, equipment IDs, timestamp accuracy, calibration status, input references, model versions, and confidence scores. These agreements ensure lineage is preserved at every handoff.

The goal is to create a balanced approach that preserves essential connections, without necessarily recording every detail. Focus on origins, major alterations, system transfers, AI interactions, and human input. This provides accountability without bogging down pipelines or systems.

Lineage also builds trust in AI. Operators can see that recommendations are based on validated inputs, proven models, and successful history, making them more likely to act on insights.

As manufacturing becomes more complex and AI powers more decisions, data lineage moves from being a nice-to-have to a core requirement. It supports regulators, empowers troubleshooting, and builds confidence in intelligent systems.

Steps to Success

Implementing comprehensive data lineage requires systematic capture and preservation of data relationships throughout the unified architecture. The steps to achieve this include:

- •**Capture Contextual Metadata at the Source** – Ensure all systems record metadata such as timestamps, locations, and event context with every data point to support complete origin tracking.

- •**Implement Lineage-Aware Data Pipelines** – Deploy tools that automatically track transformations, aggregations, and system handoffs as data moves through the architecture from edge to enterprise.

- •**Establish Model Governance and Versioning** – Maintain registries of AI model versions, training data, and configuration changes to support traceability of all automated decisions and recommendations.

- **Configure Comprehensive Audit Logging** – Capture human approvals, overrides, and actions taken based on AI recommendations within the Feedback Data Layer for complete decision tracking.

- **Deploy Metadata Management Solutions** – Implement platforms that visualize data flows, maintain relationship mappings, and provide searchable access to lineage information across the enterprise.

- **Enforce Lineage Through Data Contracts** – Define contractual requirements for metadata preservation, transformation logging, and decision documentation to ensure automatic compliance throughout data exchanges.

Data Archiving and Retention

Manufacturing data never stops flowing, but keeping every bit forever isn't practical or necessary. Smart archiving strategies preserve what matters for compliance, analytics, and continuous improvement while preventing storage costs from spiraling out of control. The key is knowing what to keep, where to store it, and how long it needs to remain accessible.

The challenge starts at the edge where sensors and controllers generate massive streams of high-frequency data. A single production line might produce millions of readings daily, but most of this raw data loses value quickly. Edge systems typically maintain rolling buffers that keep the last few hours or days of detailed readings. This provides enough history for daily troubleshooting and diagnostics without overwhelming storage capacity. Once this data gets processed through the Edge Intelligence Hub and enriched with context, the raw streams can usually be discarded.

In regulated industries, this approach shifts. Pharmaceutical and aerospace manufacturers, for example, may need to retain every sensor reading from critical processes. Here, data moves from the edge to cold storage. It's out of the way but remains accessible for audits and traceability. Data retention should also follow compliance needs, not just technical constraints.

The Common Data Model is where a solid archiving strategy really matters. This is the contextualized data consisting of the structured, labeled, and enriched information teams depend on for analysis and AI. Most manufacturers keep this data for six months to two years, depending on cycles and regulations. Older data can move to archive tiers like cloud cold storage or compressed local systems, staying searchable and retrievable when needed.

In the Unified Data Layer, the focus shifts again. The UDL handles aggregated data, not raw detail. Many teams keep this information for a few months, then transition to summary views. It's about preserving insights, not every point. Still, those insights must stay traceable back to the source. Even if raw data ages out, lineage must remain intact.

The Feedback Data Layer is different. It captures the full decision path of AI systems including predictions, overrides, actions, and results. These records are essential for audit trails, model monitoring, and continuous improvement. This data should be retained for three to five years for most industries, or even longer in regulated sectors.

Successful archiving involves having clear rules including what to keep, for how long, and in what format. Metadata must stay with the data to preserve context and traceability. Data retention policies formalize this. A policy might specify that quality records stay for seven years, while sensor readings only need 90 days of raw storage before aggregation..

Backup strategies complement archiving by ensuring business continuity. Edge systems need configuration backups and recent data snapshots to recover from failures quickly. CDM and UDL systems require more comprehensive backup approaches, including off-site replication and point in time recovery capabilities. This yields protection against both technical failures and human errors without creating excessive overhead.

Modern storage makes all of this more practical. Cloud tiers transition data automatically. Object stores scale massively with metadata for easy search. Time-series databases compress data efficiently while preserving performance. Integrated with the UMDA, these tools ensure a smooth handoff from real-time access to long-term retention.

The most effective archiving strategies balance need with cost by protecting data that drives compliance, reporting, and AI learning.

Steps to Success

Creating effective archiving and retention strategies requires balancing compliance needs, analytical value, and storage costs across the distributed architecture. The steps to achieve this include:

- **Define Tiered Retention Policies by Data Type** – Establish clear time-lines for how long different data categories remain in hot, warm, and cold storage based on regulatory requirements and business value.

- **Implement Automated Lifecycle Management** – Deploy tools that automatically transition data between storage tiers based on age, access patterns, and defined policies without manual intervention.

- **Preserve CDM Context in Archives** – Ensure that archived operational data maintains equipment, batch, and process context to support future investigations and compliance audits.

- **Establish FDL Long-Term Retention** – Create dedicated archival strategies for AI decision records, model versions, and feedback data to support explainability and continuous improvement.

- **Configure Searchable Archive Systems** – Deploy storage solutions that maintain metadata indexes and enable efficient retrieval of historical data without restoring entire datasets.

- **Formalize Data Retention Policies** – Define specific retention periods, archival formats, and compliance requirements within data contracts to ensure consistent enforcement across all data flows.

AI Optimization and Lifecycle Management

Deploying artificial intelligence is only the beginning of the AI journey. Like any essential part of manufacturing, AI needs ongoing care to remain effective across factories and enterprise platforms. Just as production equipment relies on regular maintenance to perform at its best, AI depends on consistent monitoring, timely updates, and continuous refinement. This keeps models working reliably, no matter where they're running within the architecture.

The first step in AI lifecycle management is establishing real-time performance tracking that spans the entire distributed deployment. AI models should be monitored as closely as production lines, regardless of where they operate. This means tracking prediction accuracy for edge models through the EIH, monitoring cross-domain analysis performance via the UDL, and assessing the business impact of AI outputs across all CDM domains. When accuracy starts to slip in any component of the architecture, early intervention prevents costly errors.

Fresh data is essential across the distributed environment, though sources and update needs will vary by location. Manufacturing conditions shift as new equipment, process changes, and material variations influence operations. AI models at the edge may rely on frequent updates from local sensors to stay accurate. Enterprise models benefit from periodic refreshes using broader datasets drawn through the UDL. The Feedback Data Layer helps the process of keeping everything aligned by capturing outcomes and user input to continuously guide model updates.

Scaling AI beyond pilot efforts calls for a clear, structured approach that makes full use of the distributed architecture. AI might start with focused applications in one domain, drawing on data from a specific CDM. But the greatest value comes when deployment expands to reveal cross-domain insights through the UDL. Success depends on finding the right balance between standardization and flexibility. Core frameworks can be shared across the system, while models are refined to match local conditions at each site and tailored for specific domain needs at the enterprise level.

277

LLM Routers play an important role in fine-tuning complex AI deployments across the domains and locations. They help ensure AI tasks are directed to the right model based on what's needed, whether that's query type, data source, or response time. A router might send real-time quality checks to edge-deployed models, while routing more complex cross-domain analysis to enterprise systems that access multiple CDM domains through the UDL. This step strengthens performance by minimizing latency, improving resource use, and making sure each request reaches the model best suited to deliver accurate results.

AI governance becomes more complex but increasingly important in distributed deployments. AI systems operating through the EIH must align with local operational requirements while maintaining consistency with enterprise policies. Models accessing data across CDM domains via the UDL must comply with governance policies established for each domain while supporting cross-functional decision making. This means defining what decisions different AI deployments can make, how they should handle sensitive data across domains, and how they align with broader compliance efforts spanning the organization.

Common challenges include AI model degradation occurring at different rates. Models running at the edge may drift quickly due to local condition changes, while enterprise models might degrade more slowly but affect broader operational decisions. Without proper monitoring across all deployment locations, this decline can go unnoticed until it causes real problems. The distributed nature of the architecture requires monitoring systems that can track performance across all levels simultaneously.

Version management plays a critical role when AI models run across different sites and data sources. Updates to edge models need to stay aligned with enterprise systems that are mapped to the same CDM domains. When cross-domain models change, those updates must be checked against data from multiple sources through the UDL. This careful coordination keeps the system consistent, ensuring improvements in one area don't create issues elsewhere.

Leading organizations treat AI as a constantly evolving ecosystem rather than a collection of individual solutions. They apply the same continuous improvement principles to AI that they use for production processes, but adapted for the complexity of distributed deployments. This means coordinating updates across edge and enterprise systems, ensuring consistency across CDM domains, and leveraging the FDL to capture learning opportunities from all operational areas.

Rather than pursuing a single, monolithic AI solution, organizations benefit most from developing a coordinated ecosystem of AI components that seamlessly collaborate across the distributed architecture. Success comes from focusing on real business value, maintaining models proactively, and keeping AI aligned with workforce expertise. This approach helps manufacturers build systems that grow smarter, more reliable, and more impactful with each iteration.

Steps to Success

Maintaining AI effectiveness across a manufacturing organization requires coordinated lifecycle management that spans the entire architecture. The following steps ensure AI remains a valuable, evolving tool:

- **Deploy Distributed Performance Monitoring** – Establish real-time tracking of AI accuracy, response times, and operational impact across EIH deployments and UDL accessed models.

- **Implement Coordinated Model Updates** – Create update processes that maintain consistency across edge-deployed models and enterprise systems while leveraging FDL feedback for continuous improvement.

- **Scale AI Systematically Across the Architecture** – Expand from single domain applications to cross-domain AI that leverages UDL capabilities, standardizing frameworks while customizing for local conditions.

- **Deploy LLM Routers for Insight Optimization** – Implement intelligent routing systems that direct AI queries to optimal models based on data sources, response requirements, and deployment locations.

- **Establish Distributed Testing Protocols** – Validate AI changes across local and enterprise environments before deployment to ensure coordination and prevent conflicts.

- **Maintain Comprehensive AI Documentation** – Track model versions, training data, performance metrics, and deployment locations across the unified architecture to support governance and troubleshooting.

- **Create Cross-Architecture Governance Framework** – Develop policies that ensure AI consistency across edge and enterprise deployments while maintaining domain-specific compliance.

User Enablement and Self-Service Tools

Even the most advanced architecture won't deliver value if people can't use it in their day-to-day work. For systems to make a real impact, users need to interact with AI insights and data across departments without relying on IT. The goal is to give everyone, from operators to executives, the ability to work with information directly, without needing to understand the technical details behind it.

That means self-service tools must work seamlessly across the entire system. Operators need to see local AI insights and real-time performance directly on the floor. Quality teams should explore trends across data domains without worrying about how the data is structured. Leaders need flexibility to shift from big-picture dashboards to specific metrics from individual production lines. These tools must remove roadblocks, so people can make decisions quickly and with confidence.

AI-driven discovery tools become essential for enabling non-technical users to interact with the complex, distributed data environment. Instead of navigating elaborate interfaces or understanding data source locations, users can ask natural language questions like "How did the AI optimization perform on line three today?" or "Show me quality trends across all facilities this month." These systems intelligently route queries to appropriate data sources, whether that's edge-deployed models through the EIH, cross-domain analysis via the UDL, or specific CDM repositories, returning meaningful insights without exposing architectural complexity.

Low-code and no-code platforms empower users to create custom analyses that span the distributed architecture without technical expertise. An experienced operations manager might notice patterns in local data and want to compare them with similar patterns at other facilities. With appropriate self-service tools, they can explore this hypothesis using data from multiple sources accessed through the UDL, create visualizations that combine edge and enterprise insights, and share findings with colleagues across the organization.

Effective self-service tools must handle the complexity of AI interactions across multiple deployment locations. Users need to understand when they're receiving insights from locally deployed AI models versus enterprise systems, what data sources influenced AI recommendations, and how to provide feedback that improves model performance through the FDL. The tools should make these interactions intuitive, allowing users to accept, modify, or reject AI suggestions.

Confidence in self-service capabilities builds when users understand how their actions connect to the broader architecture without needing technical details. They should feel comfortable exploring data across domains, trusting AI insights from various sources, and knowing that their feedback contributes to system improvement. This confidence comes from tools that provide clear explanations, show data lineage when relevant, and make the distributed nature of the architecture transparent but not overwhelming.

One critical consideration is making sure users can work with federated data sources without needing to understand how federation works. Self-service tools must present unified views of information, even when that data comes from multiple locations, was processed by different AI models, or spans several CDM domains. Users should be able to analyze production efficiency alongside quality metrics and supply chain data. They should do this without realizing that it involves accessing multiple different data domains through the UDL.

Context-aware assistance becomes crucial when users interact with sophisticated AI systems and distributed data sources. Self-service tools should provide relevant suggestions based on user roles, current operational conditions, and available data sources. If a maintenance technician is investigating equipment performance, the system should proactively suggest relevant historical patterns, offer AI-generated insights about similar situations, and recommend actions based on successful interventions at other locations.

When users can explore data, understand AI suggestions, and act on insights without needing help from a specialist, the system becomes a true enabler to improving how work gets done.

Steps to Success

Empowering employees to interact with data confidently and independently requires intuitive tools, role-based training, and ongoing support. The steps to achieve this are as follows:

- **Deploy Natural Language Query Interfaces** – Implement AI-driven discovery tools that allow users to ask questions in plain language and automatically route requests to appropriate data sources across the architecture.

- **Create No-Code Analytics Across Distributed Sources** – Provide intuitive tools that enable users to investigate patterns across multiple CDM domains through the UDL without exposing federation complexity.

- **Implement Context-Aware AI Assistance** – Deploy intelligent support systems that provide relevant suggestions, explanations, and recommendations based on user roles and current operational situations.

- **Establish User Friendly AI Interaction** – Create interfaces that make AI recommendations understandable and actionable while enabling workforce feedback contribution to the FDL.

- **Provide Guided Self-Service Workflows** – Implement step-by-step assistance for common analytical tasks that guide users through complex operations while building their confidence and capabilities.

Performance Monitoring and Scalability

Running a manufacturing facility without performance tracking is like operating equipment with no gauges or sensors. Teams lose visibility, and small issues can grow unnoticed. The same is true for a distributed data architecture. Without continuous monitoring across the enterprise, the system can slow down or struggle to scale as demands grow. This requires keeping a close watch on incoming and outgoing data flows to catch issues early.

Continuous monitoring across the UMDA must cover each part of the architecture while supporting the connections between them. At the site, it ensures local processing stays efficient, AI models perform accurately, and links to enterprise systems remain stable. Data federation requires tracking how information moves between domains, how distributed sources respond, and how quickly cross-domain queries are handled. Enterprise systems need checks on core infrastructure as well as enterprise AI performance and analytics that rely on combined data.

Monitoring tools must provide unified visibility across the many layers of the architecture. They also need the ability to drill down into specific components. A well-built monitoring dashboard shows real-time insights into edge processing efficiency, data flow speeds, domain health, and AI model accuracy. Integrated monitoring provides a holistic view while enabling targeted investigation when issues arise in specific areas.

Predictive analytics plays a key role in keeping distributed environments running smoothly. A slowdown in data federation response times might signal growing data volumes that could soon affect cross-domain analytics. Rising memory use on edge servers could point to the need for capacity upgrades before local performance suffers. Model drift at one facility might warn of similar issues taking shape elsewhere.

Scalability planning must account for the distributed nature of the architecture and the interdependencies between components. As manufacturing needs expand, new locations require local intelligence deployments that integrate with existing federation infrastructure. Additional data sources need integration into appropriate domains without disrupting cross-domain analytics. Growing AI capabilities must scale across edge and enterprise deployments while maintaining coordination and consistency.

A modular approach makes it possible to scale without disrupting the overall architecture. New production sites can roll out standard edge setups that connect easily to the existing federation framework. New domains can join the system to support added operations while keeping cross-domain links intact. AI can grow from local models to enterprise-wide capabilities or expand across domains without needing to rebuild the system. This flexibility supports steady growth by building on what's already in place instead of starting over.

Cloud-based infrastructure provides particular advantages for scaling distributed manufacturing architectures. Edge locations can leverage cloud resources for backup processing and data synchronization. Data federation can dynamically scale to handle increased cross-domain queries and processing requirements. Enterprise AI capabilities can expand computing resources based on analysis demands. This hybrid approach optimizes both local responsiveness and enterprise-wide scalability.

Tracking performance across a distributed architecture calls for metrics that cover both individual components and the system as a whole. At the factory location, teams need to monitor processing speed, model accuracy, and how quickly local systems connect with enterprise platforms. Federation indicators focus on query response times, data flow efficiency, and how well information moves across domains. Domain health includes checks on data quality, storage use, and how smoothly systems work together. At the enterprise level, the focus is on system availability, analytics performance across domains, and how well the architecture supports the overall business goals.

AI-powered alerting becomes essential for managing the complexity of distributed monitoring. They identify when edge problems might impact enterprise operations or when federation performance degradation could affect multiple domains. Smart alerting filters out insignificant local variations while highlighting issues that could cascade through the architecture. This enables proactive intervention before problems spread.

The aim of distributed performance monitoring and scalability planning is to keep the architecture delivering value as operations grow and change. When done correctly, monitoring runs in the background across all parts of the system. This lets teams concentrate on improving operations without getting distracted by technical limits or worrying about system failures.

As organizations add new facilities, integrate additional data sources, or expand AI capabilities, the monitoring and scalability framework should enable seamless growth while maintaining the performance, reliability, and intelligence that make the system valuable.

Steps to Success

To ensure the data architecture remains efficient and adaptable as the volume of data and use cases grow, proactive monitoring and scalability strategies must be in place. The following steps outline how to achieve this:

- **Deploy Distributed Performance Monitoring** – Implement unified monitoring that tracks edge processing efficiency, data flow performance, domain health, and AI model accuracy across all deployment locations.

- **Establish Component Specific Baselines** – Define performance benchmarks for edge processing, cross-domain analytics, data federation efficiency, and AI model accuracy that reflect the distributed nature of operations.

- **Implement Intelligent Alerting Across Architecture** – Deploy AI-powered monitoring that correlates issues across components and identifies potential downstream issues before they impact overall system performance.

- **Create Scalability Planning Framework** – Develop systematic approaches for adding new sites, expanding data domains, and scaling AI capabilities while maintaining architectural coherence.

- **Monitor Cross-Domain Integration Health** – Track the performance of data flows and analytics that span multiple domains through data federation to ensure cross-functional insights remain reliable.

- **Conduct Regular Architecture-Wide Performance Reviews** – Assess whether the system is meeting both technical performance targets and business objectives while identifying opportunities.

Insights in Action

Norman stood in the middle of the factory floor, watching as executives from headquarters toured the facility. Not long ago, this visit would have sent him into a panic. Today, he felt oddly calm.

"And this is Norman, our Lead Manufacturing Specialist," said Rita, leading the group toward him. "He's been instrumental in our digital transformation."

"Norman, the CEO would like to hear about how our new AI platform handled last week's supply chain disruption," Rita prompted.

"Well," Norman began, scratching his head as he pulled out his tablet, "when we got word that our supplier couldn't deliver on time, this fancy system went to work right away." He tapped the screen, bringing up a visualization. "It dug through all our old raw material quality records, checked prices, delivery times, and even figured out how everything would get here."

He swiped to another screen. "Then it offered three different ways we could go. We picked this option, and then orders went out, schedules got fixed, and customers got the heads-up. Simple as that."

The CEO nodded appreciatively. "How long did all that take before your... transformation?"

"Shoot, at least three days of nothing but phone calls, meetings, and a lot of aspirin," Norman chuckled, rubbing his neck. "Now it's maybe thirty minutes of us talking it through, while the AI guides us."

As the executives moved on, Rita nudged Norman's shoulder. "Remember when you used to hide when corporate visited?"

"Aw heck, that's 'cause I was always elbow-deep in breakdowns," Norman replied with a shrug. "Kinda hard to look presentable when you're sweating bullets and got grease up to your elbows."

The two of them walked past the production line where operators tapped on touchscreens, changing settings as they watched the results flow in.

"Ya know what gets me?" Norman said, scratching his chin. "I used to worry all these modern computers would push old-timers like me out the door. But turns out they just enable us to do what we've always been good at, but do it better."

"Like what?" Rita asked.

"Like last week, when I noticed that funny sound in the number four press," Norman said. "I punched it into the system and it pulled up every similar incident from the past fifteen years, showed me what fixed it and what didn't. Saved me days of troubleshooting. Or when the AI flagged those micro-vibrations in the new assembly line before anyone could hear them. Wouldn't have caught that till it broke down, before. Now we're fixing problems I never would've spotted in time, while the system's learning from our knowledge."

Their conversation was interrupted by Zoe, a recently hired production engineer seeking Norman's perspective on process variations across shifts. As Norman huddled with Zoe over her tablet, Rita watched him guide her with pride.

Later that afternoon, Norman and Rita sat in what they'd jokingly named the "War Room", once a dusty storage closet, now a collaboration space with screens displaying data from every aspect of the operation.

"There's one thing I like most about all this." Norman said, gesturing at the big dashboard with a weathered hand. "It's not the predictions or the automated tools. It's how everybody's finally talking the same language. Maintenance, production, quality, shipping, we're all working from the same playbook now."

As they walked out toward the parking lot at the end of the day, Norman paused to look back at the factory, the same buildings he'd worked in for decades, now driven with new intelligence.

"You know what's different now?" he said quietly. "For the first time in all my years, I'm not just trying to keep up with today's problems. We're actually ready for tomorrow's."

Rita smiled. "That's what this has all been about. We're not only fixing what's broken. We're building a sustainable foundation that will support generations to come."

"Well," Norman said with a proud smile, adjusting his cap, "the future better watch out, 'cause we're ready for it."

Epilogue

||||||||||||||||||||||||||||||||||||||||||||||||||||||||||||||||||||||||||||||||||||||||||||

We stand at the threshold of manufacturing's next great revolution. The convergence of data science, artificial intelligence, and industrial expertise has opened possibilities previous generations could only imagine. Manufacturing has always evolved through technological breakthroughs, from steam engines to assembly lines to computerization. Today, the transformation is driven by our ability to extract meaning and action from the vast streams of data that flow through our factories.

The concepts covered throughout this book form the foundation for this new era. When data moves freely across traditional boundaries and information is accessible to everyone who needs it, manufacturing gains new potential. This enables new tools that lead to reduced downtime, better product quality, optimized resource use, and greater flexibility to meet changing market demands.

Artificial intelligence represents the most transformative element of this revolution. AI progresses traditional manufacturing processes by detecting patterns invisible to human observers, predicting outcomes before they occur, and generating insights that would require months of analysis to discover manually. Quality systems now find microscopic defects while simultaneously identifying the upstream process variations causing them. Maintenance teams receive advance notice of equipment failures, turning crisis response into planned prevention. Supply chains reconfigure automatically to adapt to disruptions, before they impact production.

The full potential of AI in manufacturing extends far beyond these initial applications. As data architectures mature, we'll see the emergence of truly aware manufacturing systems capable of continuous learning and adaptation. Factory floors will operate as interconnected ecosystems where each process, machine, and decision point contributes to a collective intelligence. Production lines will anticipate material variations and automatically adjust parameters to maintain optimal quality. Energy consumption will dynamically optimize based on production demands, utility rates, and sustainability goals. Entire facilities will operate as learning systems, becoming more efficient with every production cycle.

These sophisticated AI capabilities depend entirely on the data foundations we've established. Consider the complexity of implementing a fully autonomous production scheduler that can adapt in real-time to changing conditions. Such a system requires access to current order status, machine availability, material characteristics, staffing levels, energy considerations, and quality parameters. Without the Unified Manufacturing Data Architecture discussed in this book, these information sources remain incompatible, making true intelligent scheduling impossible. The structured approach to data integration creates the environment where advanced AI can function effectively.

The relationship between data quality and AI capabilities strengthens as systems mature. Early AI efforts often start with pattern recognition and anomaly detection, drawing on clean, consistent data. As data governance improves, predictive models emerge, using historical trends to forecast future outcomes. When data becomes fully integrated across systems and locations, AI can shift toward prescriptive analytics, offering recommendations to fine-tune complex processes. With a strong, verified data foundation in place, autonomous systems can begin making decisions and taking action, while keeping human oversight in the loop where it matters most.

This progression illustrates how data management forms the foundation for manufacturing's AI-powered future. Those investing in these core elements position themselves to rapidly adopt increasingly sophisticated AI capabilities as they emerge. Those who neglect this foundation will find themselves unable to effectively implement even basic AI functions, falling further behind as the technology continues to advance.

The speed of learning and improvement accelerates dramatically in this environment. Insights discovered on one production line immediately inform operations throughout the facility. Solutions proven in one plant quickly scale across the enterprise. Improvements that once took months to implement now deploy in days or hours. This acceleration creates a powerful competitive advantage that compounds over time.

The integration of real-time sensor data into this framework further amplifies these capabilities. When production systems continuously monitor material properties, process conditions, and environmental factors, AI can identify previously invisible relationships between seemingly unrelated data. This deep understanding enables precise control even in highly variable manufacturing environments.

The most successful manufacturers approach this opportunity with both ambition and practicality. They envision a future where every decision is informed by data and enhanced by AI, while building toward that vision through practical steps that deliver immediate value. Some begin with quality improvement, using AI to detect emerging deviations. Others focus on asset optimization, leveraging sensor data to maximize equipment effectiveness. Supply chain visibility might be another starting point, connecting production planning with logistics and customer demand.

Each organization charts its own course based on specific challenges and opportunities, but all successful transformations share common elements of clear vision, thoughtful strategy, and commitment to execution. Technology enables this, but doesn't drive it. Leadership vision, organizational alignment, and cultural evolution provide the components for lasting change.

Human intelligence stays at the heart of this new manufacturing landscape. The most effective approaches build environments where AI supports and extends what the workforce can do. Automated systems take on routine analysis and execution, giving employees more space to focus on innovation, complex problem solving, and strategic decisions. People provide the judgment needed in uncertain situations, creativity in solving new challenges, and the context that turns data into insight. When AI and human expertise work together, they accomplish results that neither could achieve on their own.

Maintenance technicians become equipment health specialists, using AI-generated insights to prevent failures before they occur. Quality inspectors evolve into quality system trainers, helping AI models recognize emerging defect patterns. Production planners leverage AI-generated scenarios to optimize complex schedules across multiple inputs simultaneously. These enhanced roles demand new skills but build on the deep manufacturing knowledge that experienced professionals possess.

The manufacturing workforce of the future will build skills that combine deep process knowledge with digital know-how. They'll be just as comfortable working with equipment on the floor as they are interpreting the data that powers automation. This blend of expertise will only grow in importance as manufacturing systems become smarter and more connected.

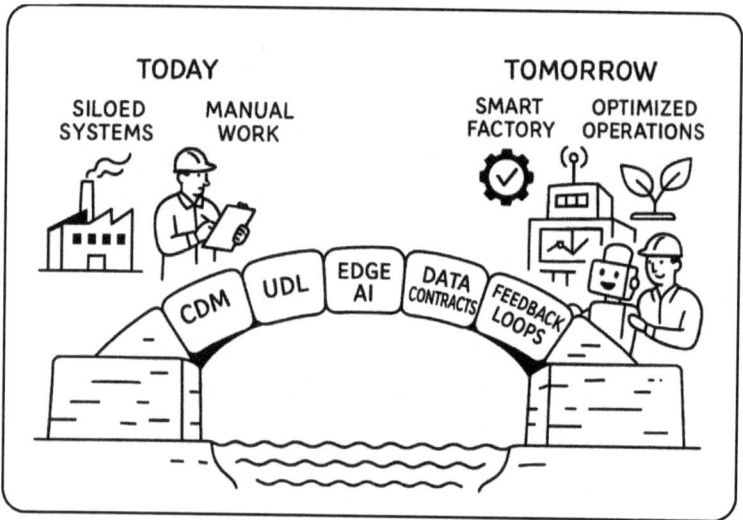

Figure E-1 Journey Across the UMDA Bridge

Looking forward, the integration of intelligent systems with physical production will continue to accelerate. Digital twins will simulate manufacturing processes with increasing fidelity, allowing virtual experimentation that optimizes operations without disrupting production. Autonomous systems will handle increasingly complex tasks, from material handling to quality inspection. AI will become more proactive, not just responding to conditions but anticipating needs and opportunities before they emerge.

The potential for multi-agent AI systems represents a particularly exciting frontier. Rather than relying on a single centralized AI, future manufacturing environments will deploy specialized agents that collaborate to achieve common goals. Maintenance agents will coordinate with production scheduling agents to optimize repair timing. Quality agents will work with process control agents to maintain product specifications while maximizing throughput. Supply chain agents will negotiate with production agents to align material deliveries with manufacturing needs. This distributed intelligence approach mirrors how human teams collaborate, with specialized expertise coming together to solve complex challenges.

Robotics and automation will evolve in parallel with these AI advances, creating physical systems capable of unprecedented flexibility and precision. Adaptive robots will reconfigure themselves for different tasks without reprogramming. Collaborative robots will work safely alongside human operators, understanding intent and adjusting their actions accordingly. Mobile robots will navigate dynamic environments autonomously, transporting materials and tools where they're needed. These physical systems will interface seamlessly with unified intelligence systems, creating truly integrated cyber-physical manufacturing environments.

The architecture that enables these advances will itself evolve. Cloud and edge computing will work in concert, processing information where it makes the most sense for each application. Real-time analytics will become standard for critical processes. Machine learning models will continuously improve, learning from each production cycle to enhance future performance. Quantum computing will eventually transform how we handle complex optimization problems that are computationally intensive for traditional systems. These technological advances will extend the capabilities of manufacturing teams, allowing them to achieve previously impossible results.

These capabilities will change how manufacturers work with both customers and partners. Product designs will evolve faster by incorporating real-world feedback and usage data. Production capacity will adjust in step with shifting demand. Supply networks will respond to actual conditions instead of relying on fixed plans. These advances will remove the barriers between design, production, delivery, and service, creating a more connected and responsive value chain across the entire product lifecycle.

Sustainability represents another frontier where data will drive impactful results. AI will optimize resource usage across multiple dimensions simultaneously, reducing energy consumption, minimizing material waste, and extending equipment lifespans. They'll enable more precise manufacturing processes that eliminate unnecessary steps and inputs. These capabilities align environmental responsibility with economic performance, turning sustainability from a compliance requirement into a competitive advantage.

Companies who thrive in this new era will combine emerging technology with human development. They'll invest in the data foundations that make AI effective and give their people the skills to use these tools with confidence. Their focus will stay on solving real business challenges, while remaining open to new ideas and innovative approaches. Most importantly, they'll foster a culture where data-driven thinking becomes second nature and continuous improvement is a part of everyday work.

What lies ahead is more than a small step forward. It is a chance to reimagine manufacturing as more efficient, more sustainable, and more rewarding for the people who make it possible. The journey there won't be easy, but it can begin today by applying the principles we've explored, and moving forward with purpose and conviction.

Appendix A - Acronyms & Key Terms

Agentic AI – A coordinated set of AI agents that can autonomously pursue manufacturing goals, collaborate, and learn from feedback.

CDC (Change Data Capture) – Technique that records and streams only the data that has changed in a source system to keep downstream stores in sync.

CDM (Common Data Model) – Standardized logical schema used to structure data consistently within a specific manufacturing domain.

CMMS (Computerized Maintenance Management System) – Platform that schedules, tracks, and records maintenance activities and asset health.

Data Catalog – Central index that documents data sources, lineage, ownership, and quality metrics across the architecture.

Data Contract – Machine-readable agreement that defines structure, semantics, SLAs, and security constraints for data exchanged between producers and consumers.

Data Historian – Time-series database optimized for high-frequency industrial process data.

Data Product – Domain-owned dataset published with a defined schema, quality SLA, lineage, and access policy so it can be reliably reused across edge, UNS, UDL, and AI layers.

Digital Thread – End-to-end connective trace of product and process data that enables lifecycle visibility.

Digital Twin – Virtual model that mirrors the state and behavior of a physical asset or processes.

Edge Computing – Placement of compute resources at or near the production equipment to enable low-latency analytics and control.

EIH (Edge Intelligence Hub) – Localized compute and orchestration layer that hosts AI models, runs real-time analytics, and synchronizes with enterprise systems.

ERP (Enterprise Resource Planning) – System for finance, planning and supply-chain coordination.

ETL (Extract, Transform, Load) – Process that integrates data from various sources, cleans and organizes it, and loads it into a central repository.

FDA (Food and Drug Administration) – Government agency responsible for ensuring that companies adhere to Current Good Manufacturing Practices.

FDL (Feedback Data Layer) – Historical repository for AI outputs, human feedback, and contextual metadata used to continuously retrain models.

GDPR (General Data Protection Regulation) – Comprehensive data privacy law enacted by the European Union.

HMI (Human-Machine Interface) – Shop floor consoles that publish operator input and display live events.

IoT (Internet of Things) – Connected sensors and controllers continuously streaming shop-floor signals.

ISA-88 (S88) – International standard that provides a framework for modeling, controlling, and managing processes in manufacturing, including standardized terminology, equipment models, and recipe management.

ISA-95 (S95) – International standard that establishes standardized structures and definitions for organizing and exchanging manufacturing data between enterprise and manufacturing systems, enabling consistent, interoperable information flow across all levels of an organization.

ISO 27001 – International standard that provides a framework for managing information security risks.

KPI (Key Performance Indicator) – Numeric measure of performance (e.g., yield, cycle-time) surfaced in real-time dashboards and aggregated reports.

LIMS (Laboratory Information Management System) – System for sample tracking and quality test results, integrated through CDMs.

LLM Router – AI orchestration service that inspects each query and routes it to the most appropriate large-language-model or analytic engine based on data locality, latency, and governance requirements.

MES (Manufacturing Execution System) – System that orchestrates production operations and gathers in-process data.

MDM (Master Data Management) – System responsible for data governance that keeps enterprise reference data (materials, suppliers, equipment) consistent across domains.

NIST 800-171 – Cybersecurity standards designed to protect sensitive government data.

OEE (Overall Equipment Effectiveness) – Composite KPI measuring manufacturing productivity as Availability × Performance × Quality.

PLM (Product Lifecycle Management) – System managing a product's design, engineering changes, and compliance through its lifecycle.

QMS (Quality Management System) – System that manages non-conformance tracking and regulatory evidence.

SCADA (Supervisory Control and Data Acquisition) – Application that aggregates PLC signals for shop floor visualization and control.

3NF (Third Normal Form) – A data normalization standard that eliminates transitive dependencies and reduces data redundancy by insuring every non-key column in a table is directly dependent on the primary key and nothing else, improving data integrity and simplifying data management.

UDL (Unified Data Layer) – Enterprise integration layer that harmonizes CDM data across domains and exposes it for cross-domain analytics.

UMDA (Unified Manufacturing Data Architecture) – Reference architecture that combines CDMs, UDL, EIH, UNS, and FDL into a scalable end-to-end data platform.

UNS (Unified Namespace) – Real-time, publish-subscribe messaging layer that organizes operational data with a hierarchical topic structure.

WMS (Warehouse Management System) – System for managing inventory including warehouse storage locations, pick-pack, and logistics.

XAI (Explainable AI) - Set of processes and methods that helps humans understand and trust the outputs of AI.

Zero Trust – Cyber-security strategy that treats every user, device, and network as untrusted until continuously verified.

Appendix B - UMDA
Implementation Checklist

Step & Key Activities	Deliverables / Milestones	Target Date
Define Vision & KPIs • Align leadership on business goals • Choose 3–5 headline metrics (e.g., OEE, first-pass yield)	Approved vision statement & KPI list	
Inventory Data Landscape • Catalogue current sources (MES, SCADA, labs, ERP) • Map data owners & quality	Data inventory register	
Prioritize Use-Cases • Score opportunities (complexity vs. value) • Select first pilots	Ranked use-case backlog	
Establish CDM Framework • Draft domain models (Production, Quality, Assets…) • Approve naming conventions • Identify domain-owned data products	CDM design docs	
Design Cross-Domain Harmonization • Define entity mappings (e.g., lot ↔ batch) • Establish shared reference datasets	Harmonization spec docs	
Deploy Unified Namespace (UNS) • Configure MQTT broker / OPC UA gateway • Publish ISA-95 topic hierarchy	UNS online & streaming events	
Build Unified Data Layer (UDL) • Stand-up storage & processing platform • Implement contextualization logic	UDL live with sample data	

Set Up Edge Intelligence Hub (EIH) • Install edge compute stack • Deploy real-time analytics rules	EIH running at first site	
Integrate Enterprise Systems • Connect ERP, PLM, CMMS, LIMS to UDL • Automate CDC or API feeds	Bi-directional data flows	
Implement Data Governance & MDM • Define data-steward roles • Establish RBAC/Zero-Trust policies • On-board master data domains	Governance policy & MDM hub	
Enable Analytics & AI Platform • Provision model development environment • Configure feature store	AI workspace ready	
Create Feedback Data Layer (FDL) • Capture AI inferences & human overrides • Store outcome metrics for retraining	FDL linked to UDL	
Execute Pilot AI Projects • Run selected use-case pilots • Measure KPI impact	Pilot tested, deployed, & lessons learned	
Scale Across Sites & Domains • Roll out UDL/UNS/EIH framework to additional plants • Expand CDM coverage	Adoption roadmap updated	
Sustain & Optimize • Establish DevOps / MLOps routines • Review KPIs quarterly • Continuous improvement loop	Quarterly KPI review deck	

About the Author

||||||||||||||||||||||||||||||||||||||||||||||||||||||||||||||||||||||||||||||||||||||

Ryan has spent more than 25 years helping manufacturers turn complex technology into real-world results. He started his career in the automotive industry, working directly with MES applications, legacy systems, lean accelerators, equipment, PLCs, and HMIs. That hands-on experience gave him a practical understanding of how factories really run, a perspective that continues to shape his work today.

Now working as a leader in manufacturing technology consulting, Ryan oversees global efforts in digital transformations involving AI and analytics enablement. His work supports some of the most advanced and demanding manufacturing environments in the world, where clear data and smart systems are key to staying competitive.

Throughout his career, Ryan has helped companies in industries including pharmaceuticals, food & beverages, industrial products, and discrete manufacturing. He specializes in applying lean manufacturing practices and his knowledge of manufacturing standards to build scalable architectures that connect legacy factory infrastructure with modern platforms in a cohesive, intelligent framework. His technical focus includes IoT enablement, data modeling, contextualized pipelines, governance frameworks, and emerging AI technologies.

What drives Ryan is a belief that data should work for the people who use it. He designs and builds architectures that are built to last, easy to maintain, and ready to grow with the evolving needs of modern manufacturing.